Foundations of Veterinary Osteology

NIPA® GENX ELECTRONIC RESOURCES & SOLUTIONS P. LTD.
New Delhi-110 034

About the Authors

Dr. Varsha Gupta is Associate Professor in the Department of Veterinary Anatomy, College of Veterinary Science and Animal Husbandry, DUVASU, Mathura and has engaged in undergraduate and postgraduate teaching since from the last 16 years. She has Published 80 research papers in various national and international journals, 85 abstracts, 02 popular articles, 02 review articles, 11 manuals, and 12 book chapters. She has published one book on "Bovine Myology: A color atlas" for the veterinary students and veterinarian. Three Mobile Apps on Veterinary Histology, Veterinary Osteology and Veterinary Myology has been developed by her. She is Reviewer of various national & international journals. She is Associate Editor of Indian Journal of Veterinary Anatomy and Editorial board member of Acta Veterinaria Veterinary Sciences. She has received IAVA Best PhD Thesis Award, many Best Paper, and Poster Presentation awards in various national and international conferences, Best M. V. Sc. Thesis guided award. She had guided two M. V. Sc. Students and member of 07 M. V. Sc. and M Sc, Biotechnology students and received best M.V.Sc. thesis guded award. She has participated in 12 National and International trainings, 4 International and 24 National conferences. She is a member of different veterinary societies like the Indian Association of Veterinary Anatomists, the Indian Poultry Science Association, Indian Society for Sheep and Goat Production and Utilization, Electron Microscope Society of India etc. The author has also organized two National Conference of IAVA, 21 days of Winter school by ICAR, webinars, and quizzes in various capacities. Her research works have been recognized at National and international levels. She is still working tirelessly for the promotion of teaching and research in Veterinary Anatomy.

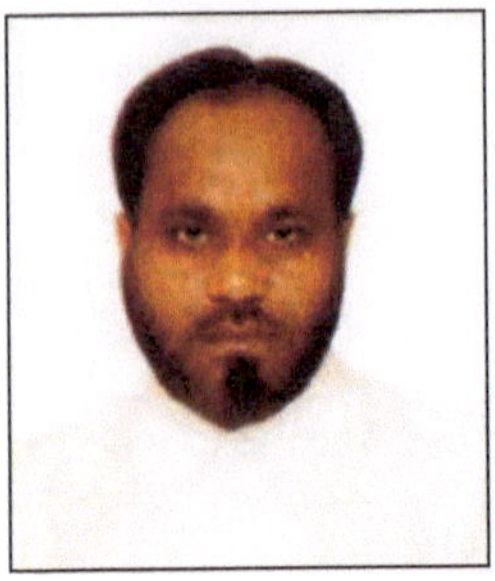

Dr. MM Farooqui is Professor and Head in the Department of Veterinary Anatomy, College of Veterinary Science and Animal Husbandry, DUVASU, Mathura and has engaged in undergraduate and postgraduate teaching from the last 35 years. He published more than 150 research papers in various national and international journals, 160 abstracts, 15 popular article, 16 manuals and 16 book chapters. He received Fellow of Indian Association of Veterinary Anatomists, Award for Best paper published in Indian Journal of Veterinary Anatomy, many best papers and poster presentation awards in various conferences, Award for best PhD thesis guided (02) and Best Teacher Award by DUVASU, Mathura. He is Life member of Indian Association of Veterinary Anatomists (IAVA). He attended more than 20 national and international symposium/conferences. He guided 03 M.V.Sc, 03 Ph.D students and act as member of advisory committee of more than 15 students. He chaired various administrative posts like Store Purchase Officer, Dean Student Welfare and Establishment officer in DUVASU, Mathura.

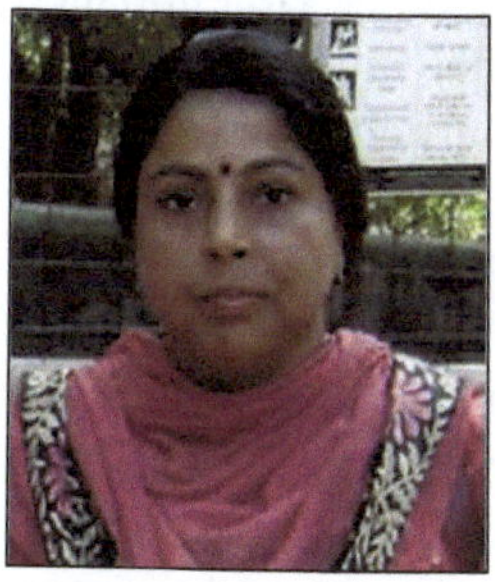

Dr. Archana Pathak is Professor, Department of Veterinary Anatomy, College of Veterinary Science and Animal Husbandry, DUVASU, Mathura and has engaged in undergraduate and postgraduate teaching from the last 25 years. She has guided three M.V.Sc. and two Ph.D theses as Major Advisor and a number of post-graduate students as a co-guide. She has published 130 research articles in the journals of national and international repute, 08 book chapters, 20 teaching/training manuals and 8 semi-technical articles. She is a distinguished Fellow of National Academy of Veterinary Science, India (NAVS-I), Fellow of Indian Association of Veterinary Anatomists (IAVA) and has been honored with several national awards such as Young Scientist Award and Gold Medal of Indian Association of Veterinary Anatomists; Award of Indian Association for the Advancement of Veterinary Research (IAAVR) for the recognition in research contributions in Veterinary Anatomy; Member of NAVS-India; Dr. P.S. Lalitha Silver Jubilee Award & Medal for the Best Paper published in Indian Journal of Veterinary Anatomy and several other Best Paper awards. She is life member of several scientific societies. Her research works have been recognized at National and international levels. She is still working tirelessly for the promotion of teaching and research in Veterinary Anatomy.

Dr. Shriprakash Singh is Associate Professor in Department of Veterinary Anatomy, College of Veterinary Science and Animal Husbandry, DUVASU, Mathura and has engaged in undergraduate and postgraduate teaching from the last 15 years. He had published 75 research papers, one book, 11manuals, 2 review articles, 27 semitechnical articles, 10 folders and 7 chapters' indifferent books. He developed a mobile app on the Veterinary Histology. He guided 2 PG students as Chairman; and member of advisory committee in 6 PG and PhD students. He organized National seminar in 2011, winter school sponsored by ICAR in 2018 as member of core committee and 2023 as coordinator. He acts as organizing secretary in User Awareness sessions on National Digital Library of India. He received NAVS (I) membership award in 2021. He also received many Best Paper and Poster Awards in various conferences. He received National level awards for the Hindi scientific article and Dr.B.V.Rao Poultry Entrepreneurs Global Icon Award by PashudhanPraharee. Dr. Singh has very successfully carried out the administrative responsibilities in the university as Hostel warden, Establishment officer, Public information officer, Recruitment officer & Incharge officer University Library.

Dr. Abhinov Verma Senior Scientist Cum Incharge Anatomy Section, ICAR-IVRI, Izatnagar, Bareilly, Uttar Pradesh and has engaged in undergraduate and postgraduate teaching from the last 10 years. He Published 62 research papers in various national and international journals, 82 abstracts, 07 popular article, 08 manuals and 07 book chapters. He received IAVA Young Scientist Award, Best Paper and Poster Presentation awards in various conferences; Best PhD Thesis Award by DUVASU, Mathura, Teaching Excellence Award by AEDS, Best Veterinarian Award by The Science World Magazine and Dr. B.V.Rao Poultry Entrepreneurs Global Icon Award by Pashudhan Praharee. He is Reviewer of various national & international journals. He is Editorial Board Member of Indian Journal of Veterinary Anatomy, HSOA Journal of Animal research and Veterinary Science, The Science World and Journal of Veterinary Reviews. He is Registered Veterinary Practitioner of UPVC affiliated with VCI. He is Life member of various scientific societies like Indian Association of Veterinary Anatomists (IAVA), Electron Microscope Society of India, ISSGAPU and Society for Immunology and Immunopathology. He attended more than 30 national and international symposium/conferences/ webinars and 16 national/international trainings. He handled various intramural as well as extramural projects.

Foundations of Veterinary Osteology
A Comparative and Applied Approach

Varsha Gupta
Associate Professor
Department of Veterinary Anatomy
College of Veterinary Science and Animal Husbandry
DUVASU, Mathura, Uttar Pradesh

MM Farooqui
Professor and Head
Department of Veterinary Anatomy
College of Veterinary Science and Animal Husbandry
DUVASU, Mathura, Uttar Pradesh

Archana Pathak
Professor
Department of Veterinary Anatomy
College of Veterinary Science and Animal Husbandry
DUVASU, Mathura, Uttar Pradesh

Shriprakash Singh
Associate Professor
Department of Veterinary Anatomy
College of Veterinary Science and Animal Husbandry
DUVASU, Mathura, Uttar Pradesh

Abhinov Verma
Senior Scientist Cum Incharge
Anatomy Section, ICAR-IVRI
Izatnagar, Bareilly, Uttar Pradesh

NIPA® GENX ELECTRONIC RESOURCES & SOLUTIONS P. LTD.
New Delhi-110 034

NIPA• GENX ELECTRONIC
RESOURCES & SOLUTIONS P. LTD.
101,103, Vikas Surya Plaza, CU Block
L.S.C.Market, Pitam Pura, New Delhi-110 034
Ph : +91 11 4386 0225, 9717133558, 9540816132
E-mail: newindiapublishingagency@gmail.com
Website: www.niparesources.com

Print ISBN: 978-93-5887-765-6

ebook ISBN: 978-93-5887-720-5

Composed and Designed by NIPA®.

Preface

The science of osteology—concerned with the structure and function of bones—holds a central position in the study of veterinary anatomy. The skeletal system provides the foundational framework upon which all other biological systems operate, influencing movement, organ protection, and structural integrity. With this in mind, *Foundations of Veterinary Osteology: A Comparative and Applied Approach* was conceived to serve as both a reference and a teaching aid for veterinary students, educators, and practitioners.

This book aims to bridge the gap between descriptive anatomy and clinical relevance by offering a comparative exploration of the skeletal systems across a variety of domestic and commonly studied animal species, including the ox, horse, dog, pig, and selected avian species. Through comparative analysis, readers gain a deeper understanding of evolutionary adaptations, species-specific variations, and the practical implications of skeletal differences in diagnosis, treatment, and animal handling.

Each chapter is organized systematically, covering the axial and appendicular skeletons with detailed illustrations, measurements, and landmarks critical for clinical and surgical applications. Where appropriate, correlations to radiographic anatomy and pathologic conditions are included to highlight the clinical significance of anatomical knowledge.

Our intent with this volume is to offer more than rote memorization of bones and their features; it is to foster a conceptual understanding of how the skeletal systems of animals relate, differ, and respond to functional demands. In doing so, we hope to cultivate a more informed and adaptable approach to veterinary practice, surgery, and research.

We hope this book serves as a valuable resource in your journey to mastering the art and science of veterinary osteology.

November 2025

Authors

Acknowledgement

We express our deepest gratitude to all those who contributed to the development and completion of this boom on Veterinary Osteology. This work would not have been possible without the guidance, encouragement, and expertise generously shared by many individuals

We extend my sincere thanks to my mentors and teachers, whose profound knowledge of veterinary anatomy and unwavering commitment to education inspired the foundation of this book. Their insightful suggestions, critical feedback, and constant support greatly enriched the scientific quality and clarity of the content.

We are grateful to our colleagues, fellow researchers, and students who provided valuable discussions, constructive comments, and motivation throughout the writing process. Their curiosity and enthusiasm for the study of animal skeletal systems served as a constant source of inspiration.

To all who contributed in ways both direct and indirect, we offer our heartfelt thanks

Authors

Contents

1

General Considerations Regarding the Bones

Veterinary Anatomy is the branch of anatomy which deals with the study of form and structure of the principle domesticated animals. The branch of anatomy which deals with the description of bones forming skeleton whose functions are support and protection of soft parts of the body is called osteology. In domestic animals' skeleton is a frame work of bony substances. The bones are covered by muscles and attached with other bone with help of ligament and cartilages to make skeleton of specific shape and form to support the body. Beside this the skeleton helps in the formation of various cavities of the body and protects the soft visceral organs present inside the cavities.

Types of Skeleton

Exoskeleton: A hard structure developed on the outside of the body is called exoskeleton. In vertebrates the term is applied to the structures produced by epidermis as hair, nails, hoof and teeth. Example in non vertebrate is the shell of a crustacean.

Endoskeleton: The bony and cartilaginous structures of the body exclusive of that part of the skeleton which is mesodermal in origin and embedded in the softer tissue form endoskeleton. The endoskeleton can be divided conveniently into three parts axial, appendicular and splanchnic skeleton.

a. **Axial skeleton:** It comprises the bones situated medial in position of the animal body viz. skull, Vertebral column, ribs and sternum.

b. **Appendicular skeleton:** It comprises the bones of limbs (forelimb and hind limb).

c. **Splanchnic/ visceral skeleton:** It consists of certain bones developed in substance of some of the viscera or soft organs, e.g. the os-penis located inside the penisof the dog, the os-cordis inside the heart of the ox, os-phrenic inside the diaphragm of camel, os-rostri inside the rostrum of pig, os-opticus in the eye of fowl etc.

Types of the bones

The bones according to their structure and function are usually divided into following classes:

Long bones: These are long, hollow, cylindrical form with extremities. The length of the bone is considerably high than the width and thickness. It comprised of a shaft/ body and two extremities. The shaft is hollow, bony cylindrical structure surrounded by medullary cavity occupied by marrow. They are weight bearing bones found exclusively in the limbs e.g. Humerus, Femur etc. In some long bones like fibula of horse, small metacarpal and small metatarsal has medullary cavity which may be very small or even absent due to arrest during development.

Short bones: The bones are comparatively small in size and generally cubical in form having almost similar dimensions viz. length, width and thickness eg. carpals and tarsals. They join together and reduce the risk of injury from concussion. Sesamoid bones which are developed in the capsule of some joint or in tendon may be included in this type. They are responsible for diminishing friction or change the direction of tendon or increase leverage to muscles and tendons.

Flat bones: It is expanded in two dimensions in excess of the third (i.e. both length and breadth are bigger than thickness) and is, therefore, more or less plate like. These provide ample area for the attachment of muscles and protects the soft organs beneath them. e.g. Scapula, cranial bones etc.

Irregular bones: These are also small bones with rough and irregular surfaces situated in the midline of the skeleton vertbrae. As these bones are situated at the middle they are unpaired bones. These bones have various processes like spinous process, transverse process etc which make them irregular. The internal constitution of these bones is similar to flat bones. e.g. Vertebrae. These bones provide extra leverage to the skeleton by attaching the strong muscles of the body.

Pneumatic bones: The long bones of the birds have cavities filled with air e.g. Humerus

Elongated bones: The long bones which do not have medullary cavity e.g. Ribs

Structure of bone: In long bone, the outer surface is covered by double membrane except at epiphysis known as periosteum. Outer layer is fibrous layer while inner is serous or osteogenic layer known as endosteum. The extremities are proximal and distal known as epiphysis or end of bones which is covered by articular cartilage whereas middle part is shaft or diaphysis. The shaft presents medullary cavity filled with bone marrow while epiphysis presents spongy substance. The compact bone is the dense and hard substance situated at the external aspect of the bone and surrounds the spongy substance.

Taxonomical position of the animals

Kingdom : Animalia

Division : Vertebrata

Class : Mammalia

Sub-class : Theria

Inter-class : Eutheria

	Cattle/ Buffalo	**Sheep**	**Goat**	**Pig/ Swine**	**Horse**	**Dog**	**Cat**
Order	Artiodactyla				Perisso-dactyla	Carnivora	
Sub-order	Ruminata			Suina	Hippo-morpha	Fissipedia	
Infra-order	Pecora			-	-	Arctoidea	Aeluroidea
Family	Bovidae			Suidae	Equidae	Canidae	Felidae
Sub-family	Bovinae	Caprinae		-	Equinae	-	-
Genus	*Bos/Bubalus*	*Ovis*	*Capra*	*Sus*	*Equus*	*Canis*	*Felis*
Species	*bycolo&indicus/ bubalus*	*aries*	*hircus*	*scrofa*	*caballus*	*Domastica*	
Sub-species	-	-	-	*Domesticus*	-	-	-

Kingdom : **Animalia**

Division : **Vertebrata**

Class : Aves

Sub-class : Neornithes

Inter-class : Neognathae

	Fowl	**Turkey**	**Duck**	**Goose**	**Pigeon**
Order	Galliformes		Anseriformes		Columbiformes
Sub-order	Galli		Anseres		-
Super-family	Phasianoidea		-		-
Family	Phasianidae	Meleagridae	Anatidae		Columbidae
Sub-family	-	-	Anatinae	Anserinae	-
Tribe	-	-	Anatini	Anserini	-
Genus	*Gallus*	*Meleagris*	*Anas*	*Anser*	*Columba*
Species	*gallus*	*gallopova*	*platyrhychos*	*answer*	*Livia*
Sub-species	*domesticus*	*gallopova*	-	-	-

Table 1: Number of bones found in the skeleton of different domestic animals and birds

Region	Cattle	Horse	Dog	Pig	Fowl
Skull including hyoid, mandible & auditory ossicles	34	34	34	30	40
Vertebral columns (Approx.)	49-51	51-57	50-53	51-56	41-42
Ribs, sternum, Coracoid and Clavicle	27	37	37	28-30 + 1	15
Forelimb	48	40	80	80	28
Hindlimb	48	40	76	80	42
Visceral bones	2 (heart)	-	1 (penis)	1 (snout)	1 (eye)
Total (Approximate)	208-210	202-208	278-281	271-278	166-167

Topographic terms

Topographic terms are used to precise the position and direction of the bone inside the body Figure 1.

Different Planes of the body

The body is divided into different planes such as median, sagittal, transverse and frontal planes.

1. **Median plane:** A plane extending through the length of the body dividing it into similar halves is known as median or medial plane.
2. **Transverse or segmental plane**: It cuts the long axis of the body perpendicular to the median plane, or an organ or limb at right angles to its long axis. A transverse plane, also known as an *axial plane* or *cross-section,* divides the body into cranial and caudal (head and tail) portions.
3. **Sagittal plane:** Planes parallel to the median plane are sagittal plane.
4. **Frontal / coronal plane**: It is perpendicular to Median and transverse planes. A coronal or frontal plane divides the body into dorsal and ventral (back and front, or posterior and anterior) portions.
5. **Longitudinal plane**: It is any plane perpendicular to the transverse plane. The coronal plane and the sagittal plane are examples of longitudinal planes.

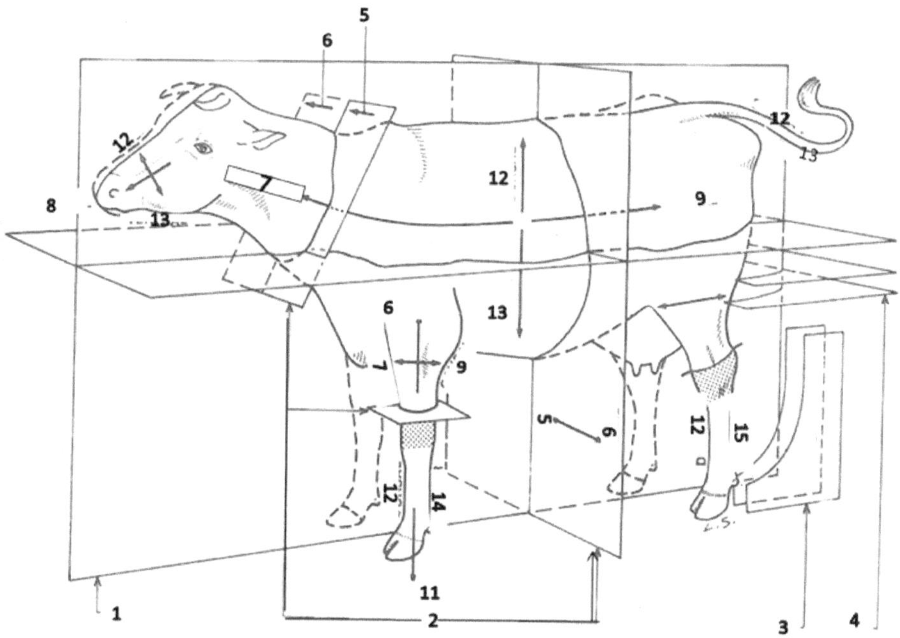

Fig. 1: Photograph Showing the Different Planes and Surfaces of the Body

1. Median plane
2. Transverse plane
3. Sagittal plane
4. Coronal plane
5. Medial surface
6. Lateral surface
7. Cranial surface
8. Rostral surface
9. Caudal surface
10. Proximal surface
11. Distal surface
12. Dorsal surface
13. Ventral surface
14. Palmar surface
15. Planter surface

Different Surfaces of a bone/ structure

1. **Dorsal:** The surface directed towards the back or sky or opposite to ventral one. Distal to the carpal and tarsal joints, the term dorsal replaces cranial surface.
2. **Ventral:**The surface directed towards the belly or ground or plane support (ground).
3. **Medial/Internal:** A structure or surface which is nearer to another median plane.
4. **Lateral/External:** A structure or surface which is away from the median plane.
5. **Anterior/Cranial/Rostral:**A structure or surface towards the head end of the body.

6. **Posterior/Caudal:**A structure or surface towards the tail end of the body.
7. **Proximal:**A structure or surface where the appendage joins the body or towards the Upper or superior
8. **Distal:**A structure or surface furthest from the point of attachment to the bodytowards or the Lower or inferior
9. **Superficial:**A structure or surface towards the close to the surface.
10. **Deep:**A structure or surface Beneath the superficial
11. **Palmar/ volar :** The surface of the fore limb that comes in contacts with the ground in standing condition. Distal to the carpal joint, the term palmar replaces caudal surface.
12. **Plantar/ solar:** The contact surface of the hind limb in standing condition below the tarsus. Distal to the tarsal joint, the term plantarreplaces caudal surface.
13. **Axial:** Towards the central line of the body or any body part.
14. **Abaxial:** Away from the axis.
15. **Oral:** The structure towards the head.
16. **Aboral:** The structure away from the head.
17. **Central:** A part nearest the middle.
18. **Peripheral:** A part nearest the surface.
19. **Parietal:**pertaining to the wall of a body cavity.The body wall or the wall of the cavity.
20. **Visceral:**associated with organs within the body's cavities. The viscera or organ in the cavity.
21. **Somatic:** The part of the body other than viscera.

Terms for description of bone

1. **Angle:** Corner of bone found between two borders.
2. **Articulation:** A joint or union between two or more bones.
3. **Body / shaft (corpus):** The largest / principal part of a bone.
4. **Border:**The edge of bone formed between two surfaces.
5. **Cornu/Hamulus:** Curved horn like process.
6. **Diaphysis:** The shaft or central portion of a long bone.
7. **Epicondyle:** A nonarticular projection in connection with or adjacent to condyle.Epicondyles is the Roughened area on the sides of a bone just proximal to the condyles.

8. **Epiphysis :** Enlarged end of a long bone. A piece of bone separated from a long bone in early life by cartilage, but later becoming a part of the longer bone.
9. **Fissure:** A narrow cleft in a bone.
10. **Fontanelle:**Un-ossified region between skull bonewhere the sutures not fully formed. The main one is between the frontal and parietal bones.
11. **Lamina:** A thin plate like structure.
12. **Linea:** Narrow crest /ridge / line i.e. A very small line.
13. **Malleolus:** A hammer headed protuberance.
14. **Manubrium:** A flat handle like projection.
15. **Neck:** A constricted portion immediately below the head.
16. **Notch:** A deep or large indentation usually for articulation.

A number of terms are used to describe the various prominences, depressions etc., of bones. Some of them are articular whereas others are non-articular i.e., gives attachment to muscles, tendons or ligaments.

Articular Prominences

1. **Head**: is a rounded articular enlargement at the end of a bone. It may be joined to the body of the long bone by a constricted part termed as 'neck'. The bony head is usually covered by articular cartilage. Examples include the femoral head and humeral head.
2. **Condyle**: Knuckle-like projections at the distal end of some long bones. Examples include the femoral and humeral condyles. They are usually covered by articular cartilage and articulate with other bones.
3. **Trochlea**:A grooved (formed by two ridges) sliding surface that acts like the surface of a pulley to increase the mechanical advantage in the movement of a joint. e.g. Trochlea of femur.
4. **Facet**: It is a small, usually flattened, articular surface e.g., costal facet.

Non-Articular Prominence

1. **Coracoid process**: is the general term for a non-articular localized elevation or prominance on a bony surface, e.g., coracoid process of scapula.
2. **Ridge**: is a non-articular extended elevation on a bony surface, e.g., trochanteric ridge of femur.
3. **Line**: is a very small non-articular low and narrow ridge, e.g., Gluteal in ilium.
4. **Spine/Spinousp process**: is a sharp non-articular a slender or pointed protection, e.g., spine of scapula.

5. **Crest:** it is a high projection or border projection of a bone. An example is the crest of the ilium.
6. **Styloid process:** Literally a pencil shaped pointed process styloid process of ulna.
7. **Tubercle:**is a small, rounded, non-articular projection which may be either smooth or rough. it is smaller than a tuberosity, e.g., teres tubercle of humerus.
8. **Tuberosity**: is a large, rounded, non-articular projection which may be either rough or smooth. It is bigger than a tubercle but smaller than a trochanter, e.g., deltoid tuberosity of humerus.
9. **Trochanter**: is the largest variety of non-articular prominance which may be smooth or rough. It is bigger than tuberosity, e.g.,trochanter major of femur.

Articular Depressions

1. **Glenoid Cavity**: is an oval, shallow articular depression e.g., glenoid cavity of scapula.
2. **Cotyloid Cavity**: is a deep, articular depression, e.g., cotyloid cavity of os coxae, cotyloid cavity of body of vertebrae

Non-Articular Depression

1. **Fossa**: It is a non-articular depression, e.g. olecranon fossa.
2. **Fovea**: It is a non-articular depression smaller than a fossa, e. g., fovea capitis of head of femur.
3. **Notch or incisura**: is a short but deep non-articular depression, e.g. acetabular notch.
4. **Groove or sulcus**:An elongated depression/concavity between parallel ridges. A furrow or intermediate between fissure and fossa. e.g. Bicipital groove of humerus.
5. **Foramen**: It is a perforation or opening in a bone for the passage ofblood vessels and nerves, e.g. nutrient foramen of femur.
6. **Pneumatic foramen**: is an opening in a bone which gives passage to air e.g., in humerus of fowl.
7. **Canal**: It is a narrow passage leading from a foramen, e.g. facial canal.
8. **Sinus:** It is an air cavity within a bone or bones; it is lined with mucous membrane and communicates with exterior, e.g. frontal sinus.

Different regions of the body and bones forming that region

Shoulder girdle/ pectoral girdle	**Scapula, coracoid and clavicle**
Shoulder	Scapula
Arm / brachium/brachi	Humerus
Fore arm/ anti brachi	Radius and ulna
Point of elbow	Olecranon process of ulna
Leg/ shin	Metacarpal bone
Manus	Carpus, metacarpus, and digits
Pes	Tarsus, metatarsus, and digits
Pastern	Ossufraginis and central carpal
Hoof	3rd pharynx
Wither	4th 5th thoracic vertebrae
Quarter	Pelvic girdle, femur and tibio-fibula
Point of hip	Tuber coxae/ coxal tuber
Hip joint	Acetabulum cavity and upper extremity of femur bone
Thigh	Femur
Gaskin and lower thigh	Tibia and fibula
Stifle joint	Distal extremity of femur and proximal extremity of tibia
Hock	All tarsal bones
Point of hock	Tuber calcis
Hamstring (Achilis tendon)	Superficial digital flexor, biceps femoris, semitendenoses and two heads of gastrochnemius
Shank/ canon	Metatarsal
Digit	1st, 2nd and 3rdphalangx
Neck	Cervical vertebrae
Thorax	Thoracic vertebrae
Loin	Lumbar vertebrae
Tail	Coccygeal vertebrae
Croup/rump	Sacral vertebra

2

Axial Skeleton

A. Comparative Study of Bones of Skull

The Skull

The skeleton of the head region is formed by the skull. The skull comprises of large number of paired or single bones. The bones of the skull get ossified with each other to form suture and with mandible form true diarthrodial joint of the skull. Most of the bones of the skull are formed by intramembranous ossification except the bones which forms the base of the skull viz. basioccipital, sphenoid, and ethmoid, developed by intra cartilaginous ossification. The bones of the skull are divided into cranial and facial groups. The bones of the cranial group form the cranium/ cranial cavity of the body which lodges the brain with its meninges, vessels and organ of hearing. The facial bones form the nasal and oral cavity and supports the larynx, pharynx and root of the tongue. The orbital cavity is formed with the help of some of the cranial and facial bones.

Cranial Bones

Paired bones		Unpaired bones	
1.	Frontal	1.	Occipital
2.	Parietal	2.	Interparietal
3.	Temporal	3.	Sphenoid
		4.	Ethmoid

Facial Bones

1. Premaxilla
2. Maxilla
3. Nasal
4. Lacrimal
5. Zygomatic
6. Palatine
7. Ptrygoid
8. Turbinates

Paired Bones

9. Vomer
8. Mandible

] Unpaired Bones

Skull of Ox As a whole, and Comparison with Those of Horse, Dog and Pig

The skull of ox is more pyramidal, shorter and relatively wider. The cranium is quadrangular and larger externally due to large frontal sinus. For description it presents four surfaces, two cavities, paranasal sinuses and three fossae.

Frontal surface: It is formed by frontal and nasal only. The frontal part is quadrilateral, extensive and have greatest width at the orbit. Cranial part presents central depression on either side for supraorbital groove and foramina. Horn cores are projected behind the median frontal eminence at lateral angle. Osseous nasal aperture is wide.

Lateral surface: It is triangular with confined temporal fossa. Temporal fossa is deep, narrow with complete boundaries separated from orbit by rounded ridge and pterygoid crest. Zygomatic arch is short, weak, flattened and formed by temporal and malar bone only. Orbital margin is completed by frontal process of malar. Pterygopalatine fossa is much larger, deeper and well defined. Sphenopalatine and maxillary foramen are deeply placed. Periorbital region is short but relatively high. Infraorbital foramen is situated above the first cheek tooth in front of facial tuberosity.

Basal surface: It is short and wide. Occipital condyles are infront of transverse ridge. Condyloid fossa presents hypoglossal and condyloid foramen. Paramastoid processes are short and converge. Foramen lacerum is slit like. External acoustic process is directed straight outward. Infratemporal fossa is small and presents foramen ovale. Posterior nares are very narrow. Hard palate is wide which forms (3/5) of entire skull length. Palatine grooves are distinct for a short distance.

Nuchal surface: It is extensive and pentagonal in outline. External occipital protuberance is located in the center for attachment of ligamentum nuchae. Median occipital crest is directed toward the foramen magnum. It is separated from frontal surface by frontal eminence and bears cornual process. Nuchal eminence is present between midway of frontal eminence and cornual process. Mastoid foramen is at the junction of occipital and temporal bone.

Cranial cavity: It is shorter and its long axis is more oblique but relatively high and wide. Ethmoidal fossa is smaller and hypophyseal fossa or sella turcica is much deeper. The deep groove extends from petrous temporal bone over foramen ovale to foramen orbito rotendum. Petrous temporal bone projects into the cavity. Temporal canal is formed entirely by temporal bone and joined by condyloid canal. Foramen lacerum is divided into two parts.

Nasal cavity: It is incompletely divided by septum. Floor is relatively long, more concave from side to side. Posterior nares are narrow and oblique. In dry skull, oval opening into palatine sinus is seen.

Paranasal sinuses: These are air spaces / cavities in the bones of cranium and the face which communicate directly or indirectly with nasal passage. In fresh state, these are lined by mucous membrane. These air spaces affect the contour of the head and add volume without weight. They protect and act as anticoncussion to vital organs.

1. **Frontal sinus:** It is the largest sinus formed by frontal bone and caudal wall of cranium. Median septum separates right and left sinus. Rostrally extend upto ventral margin of orbit, laterally upto frontal crest and supraorbital process, dorsally upto temporal fossa and root of zygomatic process and (external occipital protuberance). It is also prolonged into the cavity of the horn. It communicate with ethmoidal meatus to dorsal division of middle meatus.
2. **Maxillary sinus:** It is excavated inside the maxilla, lacrimal and malar bones. It extends upto facial tuberosity, dorsally upto orbit, caudally upto lacrimal, frontal and ethmoidal bones and nasal concha. Floor is irregular due to cheek teeth. It communicate with palatine sinus to middle meatus.
3. **Palatine sinus:** It extends from posterior border of palatine to first cheek tooth. It communicates with maxillary sinus and evacuated in palate.
4. **Sphenoid sinus:** It extends upto wing of sphenoid bone and communicate with ventral ethmoidal meatus.
5. **Lacrimal sinus:** It is excavated in lacrimal bone and communicate with maxillary sinus.

Horse: The skull is long and four sided. The foramen lacerum is extensive. The cornual process are absent. Zygomatic process meets malar as well as supraorbital process. Interparietal is centrally placed and considered as single bone. Parietal bones take part in the formation of roof of cranial cavity. Premaxilla presents six alveolar sockets for upper incisors. At the junction between maxilla and premaxilla, there is alveolus for canine tooth. The fusion between two halves of the mandible is complete. The alveolar border of the body presents six alveoli for incisor teeth. Other two alveoli are present in male for the canine tooth, which are placed further caudally. Frontal sinus is less extensive and divided into frontal and turbinate parts. Maxillary sinus is largest. Each maxillary sinus is divided by septum.

Dog: The skull is oval, elongated in shape but varies highly from breed to breed. Basioccipital join with the bulla tympani. Horn core is absent. Zygomatic process is highly curved. Interparietal fuses with occipital before birth. Parietal bones take part in the formation of roof of cranial cavity. Premaxilla has six sockets for incisors. The alveolus for canine tooth is well developed. The orbital rim is incomplete. The halves of the mandible fuses incompletely. Each half of the mandible has three alveoli for incisors, one for canine and seven for cheek teeth. Behind the angle of jaw, there is a notch and projection. Laterally, the ramus presents a depression known as masseteric fossa. There are only frontal and maxillary sinuses. Frontal sinus is well developed.

Pig: The skull is elongated in shape. Basioccipital bone is hexagonal in shape and depressed. Orbital rim is not complete. Canine teeth are well developed. Paramastoid processes are long and extended downward. Coronoid process of the mandible is short but pointed. The angle between the two rami are comparatively wide. Cranial cavity is comparatively small. Frontal sinus is subdivided by numerous septa. Maxillary sinus is small. Sphenoid sinus is very large.

Orbital fossa of Ox: It is dorsolateral depression on the surface of skull for lodgment of eyeball and its accessory structure. Each cavity is roughly conical with the apex inward. The axis of the cavity is directed downward, forward and outward. Dorsally it communicates with temporal fossa except orbital rim. It is formed by lacrimal, orbital plate of frontal, Supraorbital process, orbital rim of sphenoid and malar bones. The orbital opening of supraorbital canal is situated dorsally below which ethmoidal foramen is present. Orbital floor presents large, thin lacrimal bulla infront of which lacrimal fossa is situated. Apex of orbit is formed by optic foramen above which foramen orbitorotendum is situated.

Pterygopalatine fossa of Ox: It is the posterior extension of orbital fossa formed by palatine bone, maxilla and lacrimal bulla. It presents posterior palatine and maxillary foramen.

Horse: Lacrimal bulla is absent, Supraorbital process and frontal bone completes the dorsal rim of orbit. Lacrimal fossa is deeper. Apex presents optic foramen. Anterior to it, ethmoidal foramen and posteriorly foramen orbitale and foramen rotendum are situated. The trochlear or the pathetic foramen is placed on pterygoid crest. Behind the crest foramen alar parvum is present.

Pterygopalatine fossa is small and shallow. It does not extend upto maxilla and palatine bone.

Dog: Orbital fossa is oval. Orbital ring is incomplete as supraorbital process does not reach the zygomatic bone. Infront of optic foramen there are 2 to 3 ethmoidal foramen behind which foramen orbitale is located. Axis of orbit is more forward and less outward. Pterygopalatine fossae are continuous.

Temporal fossa: It is placed above and behind the orbit from which it is incompletely separated. It extend obliquely downward and outward on the side of cranium. It is quadrangular in outline and formed by frontal, parietal and temporal bones. It presents 3 or more foramen entering the temporal canal.

Horse: It is situated dorsolateral to cranium. It is formed by interparietal, frontal and temporal bones. 4 to 6 foramen enter the temporal cavity.

Dog: It communicates with the orbital cavity. It is formed by occipital, interparietal, parietal, frontal and temporal bones.

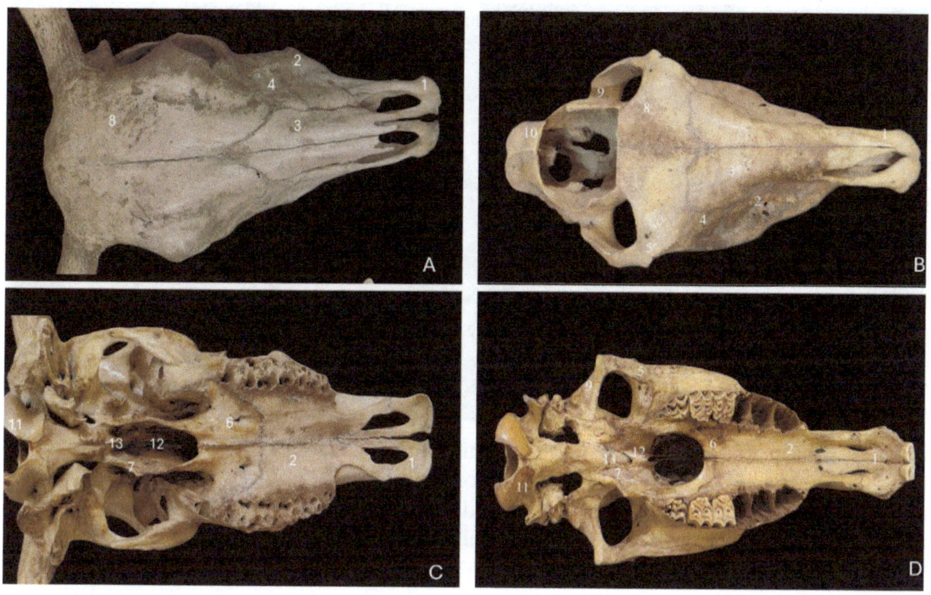

Fig. 2.1: Photograph showing the different cranial and facial bones of skull of ox and horse A. Dorsal view of skull of ox, B. Dorsal view of skull of horse, C. ventral view of skull of ox and D. Ventral view of skull of horse: 1 .Premaxilla, 2. Maxilla, 3. Nasal, 4. Lacrimal, 5. Malar/ zygomatic, 6. Palatine, 7. Pterygoid 8. Frontal, 9. Temporal, 10. Parietal, 11.Occipital, 12. Vomer, 13. Sphenoid

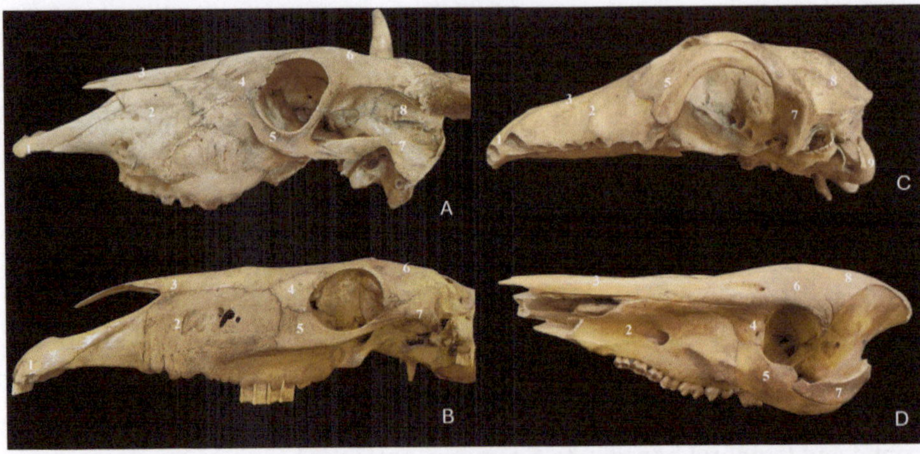

Fig. 2.2: Photograph showing the lateral view of cranial and facial bones of skull of A. Ox, B. Horse, C. Dog and D. Pig: 1.Premaxilla, 2. Maxilla, 3. Nasal, 4. Lacrimal, 5. Malar/ zygomatic, 6. Frontal, 7. Temporal, 8. Parietal, 9. Occipital

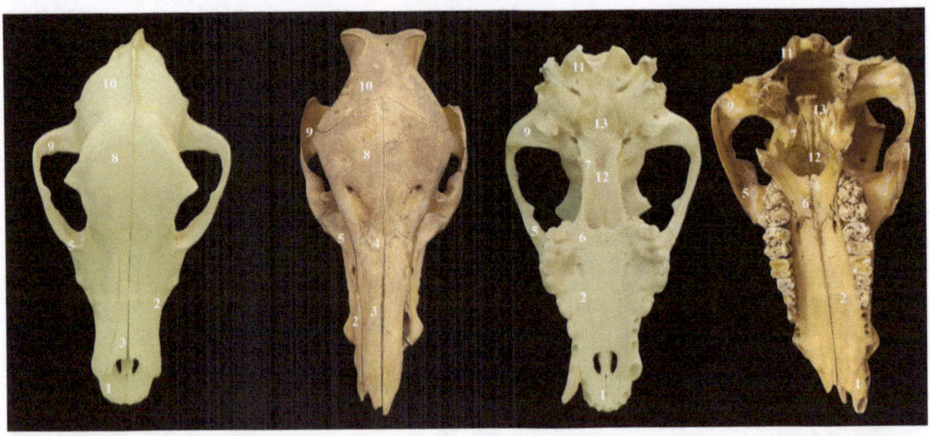

Fig. 2.3: Photograph showing the different cranial and facial bones of skull of Dog and Pig A. Dorsal view of skull of dog, B.. Dorsal view of skull of pig, C. Ventral view of skull of dog and D. Ventral view of skull of pig: 1. .Premaxilla, 2. Maxilla, 3. Nasal, 4. Lacrimal, 5. Malar/ zygomatic, 6. Palatine, 7. Pterygoid 8. Frontal, 9. Temporal, 10. Parietal, 11 .Occipital, 12. Vomer, 13. Sphenoid

a. Coparative Study of Cranial Bones of Skull

Comparative study of Occipital bone of Ox

Occipital bone is an unpaired bone situated at caudal part of cranium which form the caudal and part of ventral wall. Its lower part is centrally perforated by a large opening called foramen magnum where cranial cavity and vertebral canal join. The foramen is bounded ventrally by basilar part and dorso-laterally by lateral part of occipital bone. The bone comprises of three parts viz. lateral, basilar and squamous part (supra occipital).

a. **Lateral part:** It bears two occipital condyles placed obliquely by a wide gap dorsally and narrows distance ventrally (Fig. 2.4). Its cranial surface is concave and smooth and articular surface is convex dorso-ventrally which articulates with atlas to form occipital atlantal joint. Lateral to the condyle there is paramastoid (jugular) process (Fig. 2.4). The process is short, wide and downward and inward directed. Its medial surface is concave and lateral surface is rough for muscular attachment. Condyloid fossa presents between the root of paramastoid process and the condyle. It is in the form of a smooth depression in which two foramina are found. The ventral one is the hypoglossal foramen which transmit the 12th cranial nerve. The foramen which is located above (often double) is known as posterior or caudal condyloid foramen. A canal passes upwards from a foramen situated on the medial side of the occipital condyle and open into the temporal canal(Fig. 2.4).
b. **Basilar part:** It is strong, prismatic bar extends cranially from ventral margin of foramen magnum. It is short, wide and flattened behind and narrower and thicker in front (Fig. 2.4). Ventrally at the junction between this part of occipital bone and the sphenoid bone there are two tubercles named basilar tubercles (muscular tubercle). These tubercles give attachment to the ventral straight muscles of the head (*rectus capitis ventralis* and *longus capitis muscles*). The cranial surface of the basilar process presents a fossa which has two depressions. The cranial depression is for the lodgment of pons. The caudal depression is for lodgment of medulla oblongata (Fig. 2.6). At the junction of the basilar part of the occipital bone and sphenoid bone on the cranial side there is a crest known as spheno-occipital crest. Lateral border is thin sharp and form medial margin of pterooccipital fissure enclosing the jugular foramen. This fissure separates the jugular foramen from foramen lacerum. Jugular foramen (posterior foramen lacerum) is situated between the tympanic bulla and lateral border of basilar tubercle which gives passage to ninth, tenth and eleventh pairs of cranial nerves. An irregular opening foramen lacerum is situated between the basilar part of occipital, apex of the petrous part of temporal and sphenoid bone which passes deep and greater petrosal nerve, artery of pterygoid canal and internal carotid artery passes over it.

Squamous part: It is quadrilateral mass situated dorsal to lateral part. The supraoccipital, interparietal and parietal bones fuse before birth or soon after birth. The mass so formed by this fusion is separated from the lateral parts of the occipital bone by a transverse suture in the skull of calf. Above this suture is a central tuberosity, the external occipital protuberance, to which the ligamentum nuchae is attached (Fig. 2.4). The surface on either side of external occipital protuberance is depressed and rough for muscular attachment. There is commonly a median occipital crest which extends ventrally from the protuberance. The cerebral surface of the supraoccipital presents a central depression and above this is a variable but a pronounced eminence, the internal occipital crest is present. A groove on either side of this crest leads to the temporal canal. The mastoid foramen is situated on each side at the junction of occipital and temporal bones. It communicates with temporal and condyloid canals at the junction.

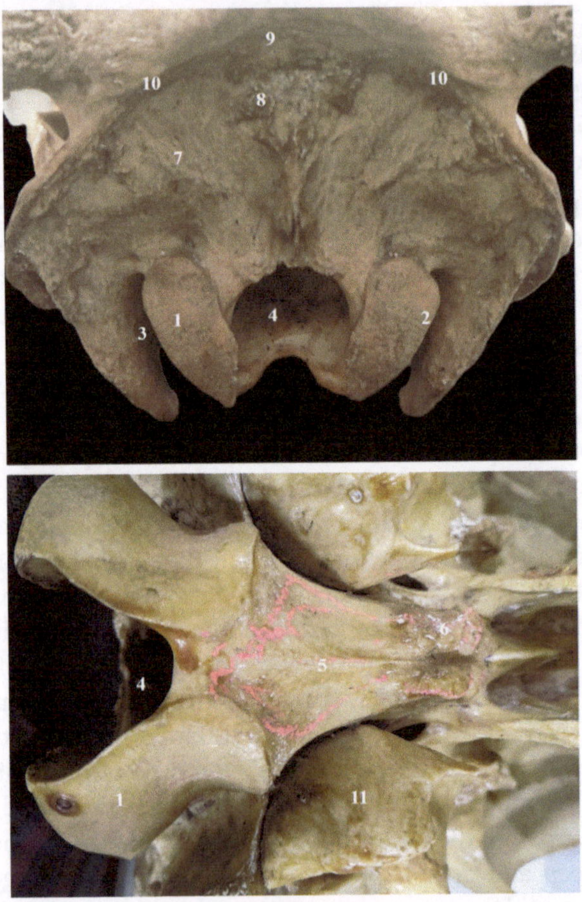

Fig. 2.4: Photograph showing the occipital bone of ox A. Caudal view, B. Ventral View: 1. Occipital condyle, 2. Paramastoid process, 3. Condyloid fossa., 4. Foramen magnum, 5. Basi occipital, 6. Basilar tubercle, 7. Squamous part of occipital, 8. External occipital protuberance, 9. Interparietal bone, 10. Parietal bone, 11. Tympanic bulla

Table 2.1: Comparative table on Occipital Bone of Ox, Horse, Dog and Pig

Characters	Ox	Horse	Dog	Pig
Participation in formation of cranium	Lower part of caudal surface of the skull	Caudal wall and a part of ventral wall.	same as in horse	Form Caudal part of the skull
Shape of foramen magnum	Wider above than below	Large and almost circular	Large , same as in ox	Trapezoid shaped
Occipital condyles	Widely apart dorsally	condyles are not so further apart dorsally	Flattened and widely separated dorsally	Basilar part also contribute in the formation of occipital condyle
Shape of paramastoid process	Short, wide and directed downward and backward and curved inward	Longer and narrower, and less curved than ox	Short, bulb shaped and do not descend below the level of condyles	extremely long and project almost straight ventrally
Condyloid fossa	Hypoglossal and one or two posterior Condyloid foramen	Contains only hypoglossal foramen	hypoglossal foramen, one or two other foramen	Hypoglossal foramen is Present
Condyloid foramen	Present	Absent	Present	Present
Internal occipital protuberance	Present in the cerebral surface of Squamous part	Same as in ox	Same as in ox	Not present
Basioccipital/ basilar part	Short and wide	Longer and prismatic, wide and flattened behind, narrow and thick cranially	Wide and joins the bulla tympanica	Hexagonal, flattened caudally
Basilar tubercle	Large and paired	Same as in ox	Present at the junction with tympanic bulla	Either side of the median ridge in the form of lateral imprints or tubercle
Jugular foramen and foramen lacerum	Short and very narrow located either side of the basilar part adjacent to tympanic bullae	Wider than ox	Same as in ox	Same as in ox
Squamous part	Quadrilateral in shape	Same as in ox	Fused with interparietal bone to form interparietal process	Triangular with base dorsally

Characters	Ox	Horse	Dog	Pig
Nuchal crest	Present but does not form highest part of skull	Prominent form highest part of skull	Prominent, angular and projects backward	Broad and prominent, form highest part of skull
Nuchal tubercle	Not present	Not present	Paired, present just below the nuchal crest	Prominent and paired
Mastoid foramen	situated on each side at the junction of occipital and temporal bones	Same as in ox	Same as in ox	Not present
Phrangeal tubercle	Not present	Not present	Not present	Present in basilar part
Tentorial process	Not present	Located in the cranial surface of squamous part which forms the osseus tentorial cerebelli	Same as in horse	Not present

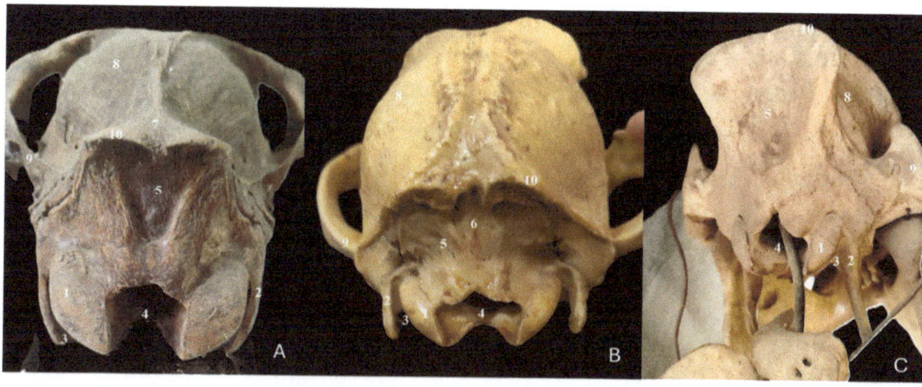

Fig. 2.5: Photograph showing the caudal view of occipital bone of A. Horse, B. Dog and C. Pig 1. Occipital condyle, 2. Paramastoid process, 3. Condyloid fossa., 4. Foramen magnum, 5. Squamous part of occipital, 6. External occipital protuberance, 7. Interparietal bone, 8. Parietal bone, 9. Temporal bone , 10 Nuchal crest

Comparative study of Sphenoid bone of Ox

Sphenoid is unpaired bone situated at the base of cranium and forms about the cranial anterior two thirds of the cranial floor. Its central part is lying infront of the basilar part of the occipital bone. It is divided into two parts viz. pre-sphenoid and post sphenoid. Each bone is consists of median body and wings laterally (Fig. 2.6)

Postsphenoid bone: The *post-sphenoid* has a *body, two temporal wings* and *two pterygoid (subsphenoidal) processes* (Fig. 2.6). The body articulates cranially with the body of presphenoid and caudally with the basi- occipital part of occipital bone. The body has two surfaces. The internal/cerebral surface presents the deep hypophyseal or pituitary fossa (*Sella turcica*) for the pituitary gland (Fig. 2.6). Bony tubercles, the posterior clinoid processes located at the superior and caudal corners of the dorsum sellae, which forms the caudal boundary of the sella turcica (hypophyseal fossa). They are variable in size and provide attachment points for the tentorium cerebelli. The external surface forms vidian groove at its junction with the pterygoid bone. The groove gives passage to the vidian nerve.

Wings (temporal): the wings are attached to the lateral border of the body and outward, forward and upward directed (Fig. 2.6, 2.7). Each wing has internal and external surface. The external surface is perforated about its middle by *foramen ovale* for passage of the mandibular nerve and middle meningeal artery. The *internal/cranial surface* presents a longitudinal groove leading to *foramen orbito rotundum.* The lateral border articulates with the squamous temporal bone. The medial border is fused with the body. The *caudal/ dorsal* *border* forms the cranial margin of the foramen lacerum. The *cranial/ ventral border* presents a notch medially which with a similar one on the wing of pre-sphenoid forms the foramen orbito-rotundum for passage of III, IV, VI cranial nerves and maxillary and ophthalmic division of V cranial nerves.

The *sub-spheniodal* or *pterygoid processes*; it is a large, thin and laterally compressed bony plate which spring from the cranial / ventral part of the wing and is directed downward and forward. It's medial surface is covered by the pterygoid and the palatine bones and lateral surface is concave. Thin cranial border of this process articulates with the palatine bone, while caudal border is free.

Presphenoid bone: It comprises of a body and pair of orbital wings (Fig. 2.6). The body is short, compressed and narrower than post sphenoid and located at a higher level, particularly in the cranial floor, than the post sphenoid. In adult, the body of pre-sphenoid is slightly excavated to form the sphenoidal sinus. The cranial/dorsal surface is flattened and is divided by a low ridge into two shallow concave areas. The cranial/ lower third part forms a raised flat area known as the jugum sphenoidale which joins the median *ethmoid spine.* At the cranial end there is a transverse depression, optic groove which supports the optic commissure in life and continues downward and forward as optic canal leading on either side to the optic foramen. The ventral/ external surface of the body is convex transversely. The anterior part of the external face of the body is concealed mostly by the vomer and laterally by the pterygoid bones.

Wings : The *orbital wings* are larger than the temporal wings. They are thick bony plates curve dorso-laterally from the sides of the body (Fig. 2.6). *External/ orbital surface* forms a part of the orbital wall. It is overlapped by the frontal in such a manner that it appears to divide into two branches. The anterior one of these joins the ethmoid and perpendicular part of palatine at the *spheno-palatine foramen.* The cranial surface is slightly concave and smooth. At its junction with the body (medial border) it is pierced by the *optic foramen* (Fig. 2.6). the lateral border is convex and articulates with the frontal bone. The ventral border also articulates with the frontal and ethmoid bones, while the dorsal border joins with the ventral border of temporal wing of post-sphenoid bone to complete the foramen orbitorotendum.

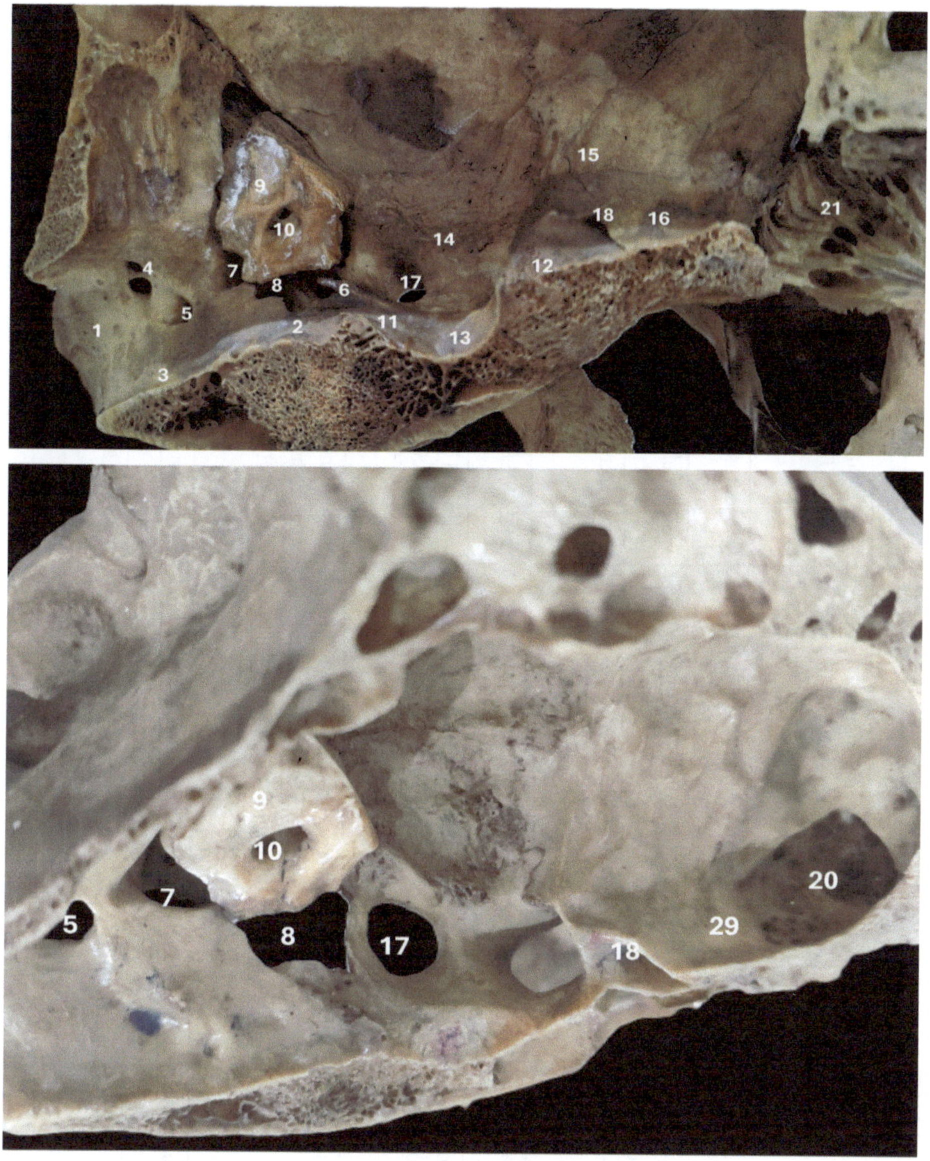

FIg. 2.6: Photograph showing the sagittal section of skull of ox

1. Basilar part of occipital bone, 2. Fossa for lodgement of pons, 3. Fossa for lodgement of medulla oblongata, 4. Caudal condyloid foramen, 5. Hypoglossal foramen, 6. Spheno-occipital crest, 7. Jugular foramen, 8. Foramen oval, 9. Petrous temporal bone, 10. Internal acoustic meatus, 11. Body of post sphenoid bone, 12. Body of the pre sphenoid bone, 13. Sella turcica, 14. Temporal wing of the post sphenoid, 15. Orbital wing of pre sphenoid, 16. Jugum sphenoidale, 17. Foramen orbitorotendum, 18. Optic foramen, 19. Ethmoid bone, 20. Ethmoid fossa, 21. Ethmoturbinates

Table 2.2: Comparative table on sphenoid bone of Ox, Horse, Dog and Pig

Characters	Ox	Horse	Dog	Pig
Post sphenoid	Shorter than horse	Placed less oblique and longer than ox body is cylindrical in shape, flattened dorso ventrally	Body is short and dorso-ventrally flattened	Resembles with ox, compressed laterally and some what triangular
Hypophysial fossa	Deep	Shallower and less distinct than ox	Deeper than horse but shallower than ox	Very deep
Dorsum sellae	Present, form caudal boundary of Sella turcica	Absent shallower and less distinct	Well developed than horse	Prominent
Tuberculum sellae	Not present	Not present	Not present	Present, form cranial boundary of Sella turcica
Post clinoid processes	present	Absent.	Well developed	Same as in ox
Temporal wing	Small and form thick prominent pterygoid crest	Irregularly quadrilateral in shape, Dorsal border presents three notches	Wide and extensive, articulates with parietals dorsally	Extend laterally and then sharply curved dorsally to articulate with temporal bone
Carotid notch	Absent	Present, formed carotid foramen by carotid notch of sphenoid and temporal bone which passes internal carotid artery and vidian nerve	Present and along with carotid notch of temporal bone form carotid foramen	Present
Oval notch	Combine form foramen oval	Passes inferior maxillary nerve	Same as in ox	Present
Spinous notch		Passes great meningeal artery		Present
Foramen oval	Present in temporal wing	Absent	Placed near the dorsal border of temporal wing	Absent

Characters	Ox	Horse	Dog	Pig
Foramen rotundum	Combined to form foramen orbit rotundum	Present near the cranial border of temporal wing	Opens into alar canal which pierce the root of pterygoid process	Similar to ox
Foramen orbital		Present at the junction of orbital and temporal wing	at the junction of the wings a little lower to the optic foramen.	
Small trochlear foramen		may be present on pterygoid crest	Absent	
Alar canal	Absent	Perforate the root of pterygoid process which transmit maxillary artery and opens into small alar foramen	Present and foramen rotendum opens into it	Absent
Alar foramen	Not present	Small	Two in number cranial and smaller caudal foramen	Absent
Presphenoid	Body is short	Same as in ox	Body dorso-ventrally flattened	Body short and compressed laterally
Sphenoidal sinus	Absent in calf and small in adult	Continue with the palatine bone	Absent	Very large
Pterygoid fossa	Not present	Not present	Not present	Broad and twisted pterygoid bone form pterygoid fossa with

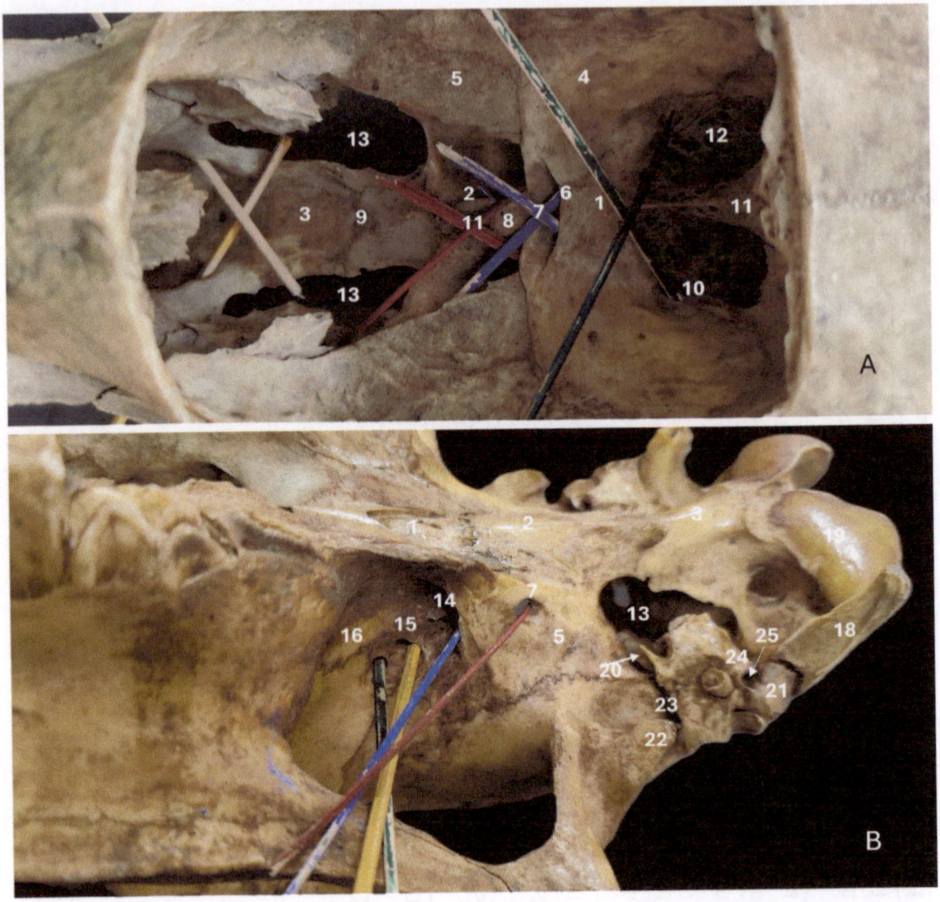

Fig. 2.7: Photograph showing the skull of Horse A. dorsal view after removal of part of frontal and parietal bones, B. Ventro¬lateral View

1. Body of presphenoid
2. Body of post sphenoid bone
3. Basilar part of occipital bone
4. Orbital wing of pre sphenoid
5. Temporal wing of the post sphenoid
6. Jugum sphenoidale
7. Optic foramen
8. Hypophyseal fossa
9. spheno-occipitsl crest
10. Ethmoid Foramen
11. Crista galli
12. Ethmoid fossa
13. Foramen lacerum
14. Foramen orbital
15. Foramen rotundum
16. Ethmoid foramen
17. External acaustic meatus
18. Paramastoid process
19. Occipital condyle
20. Muscular process
21. Mastoid process
22. Retroauricular process
23. Retroauricular foramen
24. Styloid process
25. Stylomastoid foramen

Comparative study of Ethmoid bone of Ox

Ethmoid is unpaired bone lies infront of the body and orbital wing of presphenoid which forms the anterior portion of floor of cranial cavity. It projects forward between the orbital parts of the frontal bones and enters into the formation of the cranial, nasal and paranasal cavities. It consists of four pats viz. cribriform plate, two lateral masses or labyrinth and perpendicular plate. The cribiform plate (lamina cribrosa) is placed transversely at the ventral end of cranium which is sieve like partition between the cranial and nasal cavities (Fig. 2.8). The sieved like character of this plate is due to the presence of numerous small (olfactory) foramina for the passage of the olfactory nerve filaments. Its margin joins laterally with the orbital wings of the sphenoid, ventrally with the body of the presphenoid and dorsally with the cranial plate of the frontal bone. The intracranial projection of the perpendicular plate divides its cerebral surface into two parts by a median ridge called the crista galli (Fig. 2.8). Each half form a deep oval large depression called ethmoidal fossae, which lodges the olfactory bulb. A large ethmoidal foramen is present on either side for the passage of the ethmoidal vessel and nerves. Perpendicular plate is extensive situated in the median plane and forms the basal part of the osseum septum nasi. Its lateral surfaces (right and left) are nearly plain. They are covered by nasal mucous membrane. They are separated by a narrow interval from the lateral masses and form part of the inner wall of the nasal chamber. Its cranial border is irregular and is continuous with the septal cartilage. The caudal border projects into the cranial cavity as a ridge, the crista galli. It becomes continuous with the median crest of the frontal above and the ethmoidal spine of the presphenoid below. The dorsal border joins the frontal bones at their line of junction. The ventral border is received into the groove of the vomer. Lateral masses consist of large number of delicate scroll like plates termed ethmo-turbinates (Fig. 2.8). Each lateral mass or labyrinth is a conical mass with its base attached to the lower part of cribriform plate. It projects forward into the caudal part of the nasal cavity. The medial surface is separated from the perpendicular plate by a narrow space. The lateral surface is convex and faces chiefly into the frontal and maxillary sinus. It is covered by a thin layer of bone, the lamina lateralis which gives attachment to ethmo terbinates. The ethmo turbinates are separated by narrow intervals termed the ethmoidal meatuses, which communicate with the nasal cavity. In the fresh state the ethmo turbinates are covered with mucous membrane. The lateral mass consists of five endoturbinates and eighteen ectoturbinates. The largest ethmoturbinate is so extensive as to be termed a third or middle turbinate bone. It projects forward between the dorsal and ventral turbinates.

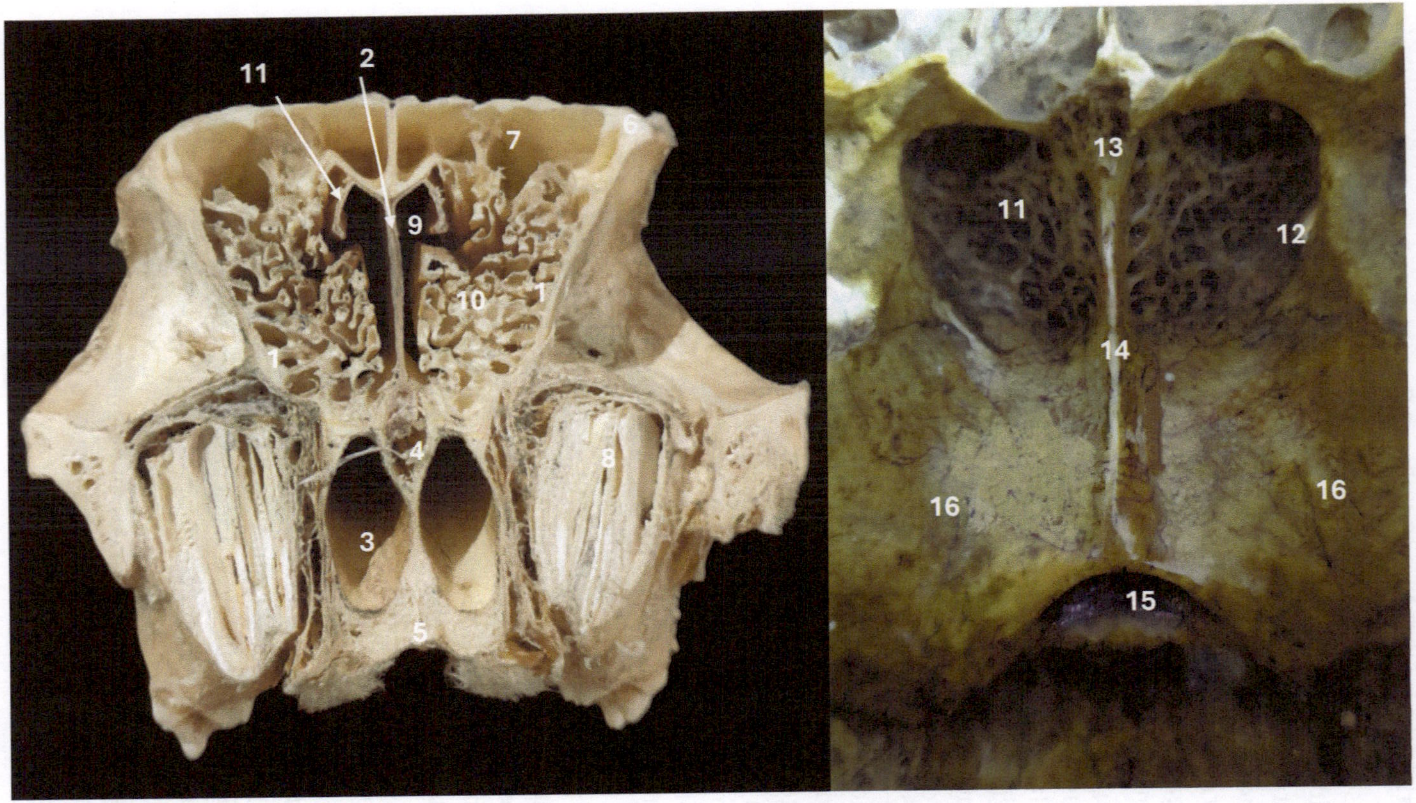

Fig. 2.8: Photograph showing the transverse section of skull of Ox (at the region of ethmoid bone)
1. Lateral mass of ethmoid bone, 2. Perpendicular plate of ethmoid, 3. palatine sinus, 4. Vomer, 5. Palatine bone 6. Frontal bone, 7. Frontal sinus, 8. Cheektooth, 9. Common nasal meatus, 10 Ethmo turbinates 11. Cribriform plate 12. Ethmoid foramen, 13. Crista galli, 14. Ethmoidal spine of sphenoid bone, 15. Optic canal, 16. Orbital wing of sphenoid, 17. Dorsal nasal concha

Table 2.3: Comparative table on Ethmoid bone of ox, horse, dog and pig

Characters	Ox	Horse	Dog	Pig
Cribriform plate	Developed and grooved	Narrow and deeply grooved than ox	Very extensive than ox	More developed than ox
Ethmoidal/ Olfactory fossae	Deep	Smaller than ox but deeper and oval in shape than ox	Very deep	Same as in ox
Arrangement of ethmoturbinates	2 rows (both ecto and endo turbinates in single row each)	Three rows (endoturbinate,-single row, ectoturbinate-double row)	Two rows (both ecto and endo turbinates in single row each)	2 rows (both ecto and endo turbinates in single row each)
Endoturbinate	Four	Six	four	seven
Ectoturbinates	18	21	Six	20
Crista galli	Developed	As ox	Faint ridge form of incomplete	Well developed, same as ox
Perpendicular plate	Extensive, form the caudo-dorsal part of the nasal septum.	Less extensive than ox, Forms the upper two-thirds of the nasal septum.	Longer than ox	Very long than ox, horse and dog
Orbital plate	delicate, paper-thin plate of bone that forms dorsal margin of sphenopalatine foramen	Forms the inner wall of the eye socket. It separates the orbit from the numerous ethmoidal air cells (sinuses) within the ethmoid bone and provides a structural barrier between the orbital cavity and the nasal cavity.	Extensive, form medial wall of maxillary sinus	Forms part of the medial wall of the pterygopalatine fossa

Characters	Ox	Horse	Dog	Pig
Lateral mass	cone-shaped and highly developed, with extensive scroll-like ethmoturbinates that project further into the nasal cavity.	More delicate than ox and inclined towards the ventral end	Well developed and Projected into frontal sinus	relatively less developed and do not extend as far forward into the nasal cavity.
Middle nasal turbinate	Present	Not formed	Same as in ox	small and projects only a short distance from the ethmoidal labyrinth.
Lamina lateralis	a thin, curved, lateral plate of bone that covers each of the ethmoidal labyrinths	Same as in ox	an extensive part of the ethmoid bone that forms the medial wall of the maxillary sinus.	Same as in ox
Lamina transversalis	Not present in ox	Not present in horse	a shelf extending inward from the lower part of the lamina lateralis of the ethmoid bone which joins with a similarly curved part of the palatine bone and out-curving plates of the vomer. It separates the upper part of nasal cavity from nasal meatus	A smaller, less prominent structure that is part of the ethmoidal labyrinth. It gives rise to dorsal and ventral spiral lamellae.it Is involved in creating the ethmoturbinates but does not form the same distinct division as in the dog.

A linear series of relatively large foramen are found on either side of crista.

Comparative study of Interparietal bone of ox

Interparietal bone is paired bone situated centrally between squamous part of occipital and parietal bone. It presents external parietal and internal cerebral surface and lateral, caudal and rosrtal borders. It has no intercranial projection and tentorial process.

Table 2.4: Comparative table on Interparietal bone of ox, horse, dog and pig

<table>
<tr><th>Characters</th><th>Ox</th><th>Horse</th><th>Dog</th><th>Pig</th></tr>
<tr><td>Fusion</td><td>Fused with occipital and parietal bones before birth</td><td>Fused with occipital and parietal bones in adult stage (variable)</td><td>Fused with supra occipital bone</td><td rowspan="5">Fused with squamous part occipital bone during development or absent</td></tr>
<tr><td>Location</td><td>Wedged between Occipital and parietal bones</td><td>Same as in ox</td><td>Wedged between Parietal bones</td></tr>
<tr><td>Rostral end/ border</td><td>Thin</td><td>Same as in ox. Give attachment to falx cerebri</td><td>Narrow and thin</td></tr>
<tr><td>Caudal border</td><td>Thick</td><td>Same as in ox</td><td>Same as in ox</td></tr>
<tr><td>Internal / cerebral surface</td><td>Process absent</td><td>Three sided process projects between cerebrum and cerebellum</td><td>Forms osseous cerebellar tentorium</td></tr>
</table>

Comparative study of Parietal bone of ox

The parietal bones are fused before or soon after birth, with the interparietal and the supra occipital. The bone does not participate in the formation of the roof of cranium. It constitutes the upper (parietal plane) and lateral wall (temporal plane) (Fig. 2.2, 2.4). The parietal plane forms the part of the caudal wall and temporal plane forms the lateral boundary of temporal fossa. The line of inflection between two planes is marked by the prominent parietal crest, which continue cranially with the frontal crest and below with the temporal crest. These bones facilitate the frontal bones in the formation of central frontal eminence which is highest point of skull. The parietals are excavated to form the part of frontal sinus in the adult animals.

Table 2.5: Comparative table on parietal Bone of ox, horse, dog and pig

Characters	Ox	Horse	Dog	Pig
Relation with other bones	Fuses before or soon after birth, with the interparietal and the supra occipital	Quadrilateral plate of curved bones with little cancellated tissue between two compact layers	Rhomboid in outline and is strongly curved	Overlapped by occipital bone
Participation in roof of the cranium formation	Do not participate	Participate in the formation of roof of cranium	Greater part of the roof of the cranial cavity	Participate in the formation of nuchal crest
Parietal crest	Prominent	Continuous with the inter-parietal and occipital and frontal crest	Prominent, continued upon the frontal bones	Absent
Temporal line	Absent	Absent	Absent	Present between parietal plane and temporal plane
Parieto temporal canal	Formed by the junction of the parietal and temporal bones	Posterior border meets with occipital and temporal bone to form the parieto temporal canal	Formed by the junction of the parietal and temporal bones	Absent or poorly developed
Internal parietal crest	Present internally in the form of prominent ridge	Same as in ox	Same as in ox	Same as in ox

Comparative Study of Frontal Bone of Ox

Frontal bone is paired bilateral symmetrical, very extensive, bone situated between parietal and nasal bone. It forms about one half of the entire length of the skull and complete the roof of the cranium. It articulates with ethmoid, lacrimal, wing of presphenoid, palatine, maxilla and squamous part of zygomatic process of temporal bones. It presents four parts *viz* squamous, nasal, orbital and temporal parts (Fig. 2.2, 2.9).

The Squamous part/ body form the base of the forehead. It has two surfaces and four borders. External surface is convex for the most part, smooth and is covered by the thick *frontalis* muscle. Supraorbital foramen (often double) is situated about an inch medially from the root of the supra orbital process pierces this surface and continue in the supra orbital groove (lodges frontal vein). The supra orbital canal extends from supraorbital foramen which passes down ward and forward to the orbit. The Internal surface forms cranial and nasal cavity and consists of two plates cranial and nasal (Fig 2.10). Cranial plate is concave and marked by digital impressions. It has internal frontal crest in which falx cerebri of duramater is attached. The nasal plate runs downward and forward to form the roof of the nasal cavity. External and internal plate encloses large intervening frontal sinus (Fig 2.10). The medial border joins its fellow at the frontal suture. Lateral border is divided into three parts *viz.* temporal, orbital and lacrimal. The Temporal part is concave and form cranial boundary of temporal fossa (Fig 2.10). The orbital part is thin and serrated to form cranial rim of orbital fossa. Lacrimal part overlaps the lacrimal bone. The ventral/ cranial border slopes outward and articulates with nasal bone at naso-frontal suture. The ventral borders of two frontal bones meet with each other to form a notch which receives the nasal bones. The Dorsal/ caudal border articulates with parietal bone at parieto-frontal suture and form a large central frontal eminence or torus frontalis i.e. the *highest point of the skull* with the parietals (Fig 2.10).

The horn core or Cornual process is a hollow conical osseous structure projects from its dorsolateral angle (Fig 2.10). It is elongated conical form, and varies greatly in size, length, curvature and direction. Its outer surface is rough and porous marked by numerous grooves and foramina; in the fresh state it is covered by the corium of the horn. Internally it is divided by bony plates and communicate with frontal sinus.,

Supraorbital process forms the part of the orbital rim (Fig 2.10). It curves outward and downward from the body partially separates the orbit from the temporal fossa. Its upper surface is convex while the orbital surface is concave and smooth for the lodgment of the lacrimal gland. The lower edge is thin and concave and forms part of the upper rim of the orbit. The upper edge is slightly concave and thicker than the lower edge. The outer extremity is nearly flat and meets the zygomatic process of malar.

Nasal part is pointed and wedge between the nasal bones and lacrimal bone (Fig 2.10). Zygomatic process forms the bridge between squamous part and temporal bone. This process partially separates the orbit from temporal fossa and supraorbital foramen is situated at this point.

Orbital part/ Orbito-temporal plate is concave and smooth to form greater part of the orbital wall and has opening of supra orbital canal. Ethmoid foramen is located close to its posterior border.

Temporal plate is small and concave. It is separated from the orbital part by sphenoid notch.

Table 2.6: Comparative table on Frontal Bone of Ox, Horse, Dog and Pig

Characters	Ox	Horse	Dog	Pig
Participation in cranium roof formation	very extensive, bone complete the roof of the cranium	less extensive than ox, forming only a part of the roof of the cranium.	Less extensive than ox and horse, irregular in shape, being broad caudally and somewhat narrower, forming only a part of the roof of the cranium Rostrally	Irregular in shape, forming only a part of the roof of the cranium
Cornual process	Present at the junction of lateral border of frontal and parietal	Absent	Absent	Absent
Inter-cornual protuberance	Forms the highest part of skull by parietal border of frontal with parietals	Does not formed by frontal but by squamous and lateral parts of the occipital bone.	formed by the occipital bone and interparietal bone.	absent
Supra orbital process	Joins the zygomatic process of the squamous temporal and malar	Long and curved, joins the zygomatic process of the squamous temporal only	Short and blunt and the orbital rim is incomplete	Short, blunt, pointed and is not connected with the zygomatic arch

Characters	Ox	Horse	Dog	Pig
Supra orbital foramen	Situated about 2.5 cm medial to the root of supraorbital process	Perforate the root of the supra orbital process or may be represented by a notch on its anterior border.	Absent	Placed in between the orbital margin and frontal suture
Supra orbital groove	Continue Rostrally from supraorbital foramen	Absent	Same as in ox	Continue Rostrally from supraorbital foramen onto nasal bone
Ethmoid foramen	Perforate the orbital wing near its lower border Open above the supraorbital canal	Situated above the optic canal at the junction of the orbital part of the frontal and the orbital wing of sphenoid.	Two in number , approximately 1 cm rostral to the optic canal.	Same as in ox
Sphenoid notch	Present and receive the wing of pre sphenoid bones.	Same as in ox	Absent	Present and receive the wing of pre and basi sphenoid bones.
Articulation with palatine bone	Does not articulate with palatine bone	Laterally unite with the palatine bone also	Articulates with the palatine bone to form the fronto-palatine suture	Does not articulate with palatine bone
Nasal articulation	Nasofrontal part form notch which receives the nasal bone	United lower ends of the two frontals form a projection which fits into the notch formed by the united upper extremities of nasal bone	There is a narrow nasal part, which fits in between the nasal bone and maxilla	Small rostral part fused with nasal bone at nasal border
Ethmoid fossa	Contribute in formation of ethmoid fossa	Contribute in formation of ethmoid fossa	Contribute in formation of ethmoid fossa	Contribute in formation of ethmoid fossa
Temporal part	Extensive	Less extensive than ox	Less extensive than ox and horse	Less extensive than ox but more extensive than horse

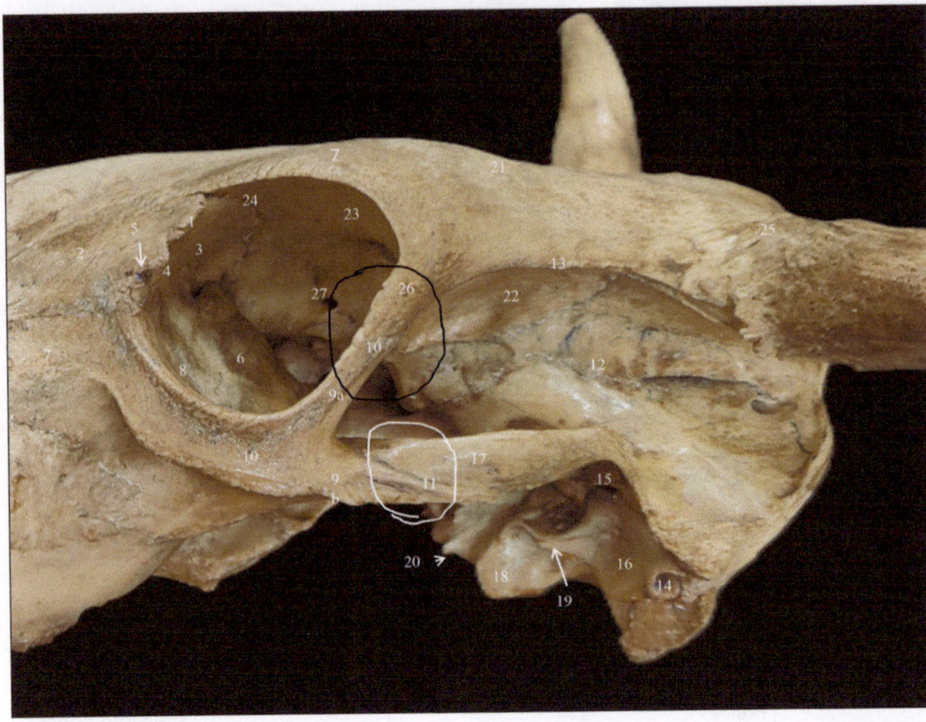

Fig. 2.9: Photograph Showing the Lacrimal, Malar, Temporal and Frontal bones of skull of Ox

Lacrymal

1. Orbital margin of lacrimal
2. Facial surface of lacrimal
3. Orbital surface of lacrimal
4. Lacrimal fossa
5. Lacrimal foramen
6. Lacrimal bulla

Malar

7. Facial surface
8. orbital surface
9. Zygomatic process of malar
   a. Frontal part
   b. Temporal part
10. Supra orbital arch
11. Zygomatic arch

Temporal

12. Temporal fossa
13. Temporal crest
14. External acoustic meatus
15. Post glenoid foramen
16. External auditory process
17. Zygomatic process of temporal
18. Tempanic bulla
19. Hyoid process
20. Muscular process

Frontal

21. Squamous part
22. Temporal part
23. Orbital part
24. Lacrymal part
25. Cornual process
26. Supra orbital process of frontal
27. Ethymoid foramen

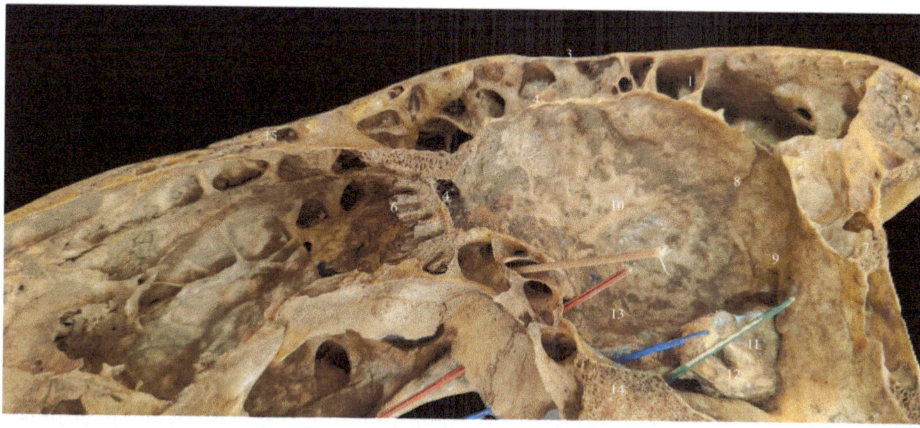

Fig. 2.10: Photograph Showing the sagittal section of skull of Ox

1. Frontal sinus
2. Frontal eminence
3. External plateof frontal bone
4. Internal plate of frontal bone
5. Ethymoid fossa
6. Ethmoterbinates
7. External occipital protuberance
8. Parietal bone
9. occipital
10. Cranial cavity
11. Petrous part of temporal bone
12. Internal acaustic meatus
13. Temporal bone
14. Sphenoid bone
15. Nasal plate of frontal bone

Comparative Study of Temporal Bone of Ox

Temporal bone is a paired bone which forms part of the lateral wall of cranium and situated between the occipital bone above, parietal infront and malar below, frontal dorsally and sphenoid ventrally and medially (Fig. 2.9). It consists of squamous and petrous parts which are fused completely at birth.

Squamous temporal part consists of body and zygomatic process. Body is irregularly oval curved plate presenting two surfaces and four borders. Temporal crest is the downward continuation of parietal crest which turns forward below ending in a tubercle, above the external acoustic meatus. The crest divides the body into two parts. The area in front of the crest is known as *Temporal fossa* and it presents two or three foramina directing into the temporal canal. The area behind the crest is smaller and congregates the occipital bone (Fig. 2.9). The *internal/ cerebral surface* is overlapped by the parietal and basisphenoid bones. The external/ temporal surface is convex and helps in the formation of temporal fossa. The zygomatic process arises from ventral part of temporal fossa. It is wider behind and narrower in front and forms the lateral boundary of temporal fossa. It is in the form of a long, flat, curved plate directed outwards, downwards and forwards to meet the ventral plate of zygomatic process of the malar to form zygomatic arch (Fig. 2.9).

An articular tubercle (temporal condyle) is located ventral to the zygomatic process for articulation with the mandible. Behind the condyle mandibular fossa (*a glenoid cavity)* and retroauricular process (*post-glenoid process)* is located. Retroauricular foramen (*Post glenoid foramen)* is present behind the post-glenoid process which acts as an external opening of the temporal canal. The dorsal (parietal) border of squamous part articulates with parietal bone forming squamous suture. The ventral (sphenoid) border join the temporal wing of basisphenoid. The rostral (frontal) border articulate with frontal bone and caudal border articulate with occipital and petrous part of temporal bone (Fig. 2.9).

Petrous temporal part is situated on the caudolateral aspect of the cranium between the occipital behind and the squamous temporal in front. It consists of *petrous* and *tympanic parts*. The *petrous part* is small and hardest bone in the skeleton which contain internal ear. The medial surface is smooth and forms the lateral wall of the cerebellar compartment of the cranial cavity. The internal acoustic (auditory) meatus is present on the medial surface of petrous (for transmission of VII and VIII cranial nerves) (Fig. 2.10). A crest, petrosal crest separates the rostral and medial surface. The caudal (occipital) surface is slightly concave. The Lateral surface is concealed by tympanic bulla. A narrow fissure, petro-occipital fissure separated the lateral and ventral surface.

Tympanic part is extensive and comprises of external auditory process, auditory/ tympanic bulla and muscular and styloid (hyoid) processes. The *external auditory process* is a curved plate of bone projecting through a notch in the squamous temporal and encloses the external acoustic meatus. The *styloid process* is located below the external auditory process. This is connected by a bar of cartilage to the styloid cornu of the hyoid bone. Stylomastoid foramen is situated above the process between the styloid and the paramastoid processes which is external opening of the facial canal. The cavity of *bulla tympanica* (auditory bulla) forms part of the middle ear. The *muscular process* of the petrous temporal springs from below the bulla for muscular attachment. It is large and its free end is often bifid (Fig. 2.9).

Table 2.7: Comparative table on Temporal bone of Ox, Horse, Dog and Pig

Characters	Ox	Horse	Dog	Pig
zygomatic process	Shorter and weaker than horse articulates with malar only	meets with zygomatic process of malar and maxilla and frontal bone	strongly curved, articulates with zygomatic process of malar	Same as ox
Retroauricular tubercle	Convex from medial to lateral and cranial to caudal direction	Transversely elongated	absent	Same as in horse
retroarticular process	Less prominent	More prominent	Well developed than ox and horse	Absent
External acoustic meatus	Smaller than horse, directed laterally	Larger than ox	very short and wide	Elongated and narrow
Bulla tympanica	Large and laterally compressed	poorly developed	very large and rounded and smooth, its medial side is united to the basilar part of the occipital bone	Elongated and compressed laterally
Petrosal crest	indistinct	Obliquely caudally directed crest which separated medial and cranial surface of petrous temporal	sharp and prominently projects into the cranial cavity	Absent instead of that tentorial crest is present
Petro occipital canal	Present in the form of fissure	Present in the form of wide fissure	Located between petrous temporal and basal part of occipital bone	In the form of narrow fissure
Styloid process	Cylindrical in shape , project downward -forward, gives attachment to hyoid apparatus	Same as in ox	Rudimentary/ absent	Oval, short slim rod shaped

Characters	Ox	Horse	Dog	Pig
Mastoid process	Absent	well developed, bulb shaped, present between the squamous-temporal and the paramastoid process of occipital bone.	Distinct , gives attachment to hyoid apparatus	Reduced
Stylo mastoid foramen	Present between styloid and paramastoid process	Present between styloid and mastoid process	Situated between mastoid process and tympanum	Same as in ox but incompletely separated from styloid process by a bony shelf
Muscular process	Large and often bifid at its free end	Smaller than ox	rudimentary	rudimentary
Nuchal process	Absent	Absent	Absent	Present between external acoustic meatus and occipital bone, gives attachment to hyoid apparatus

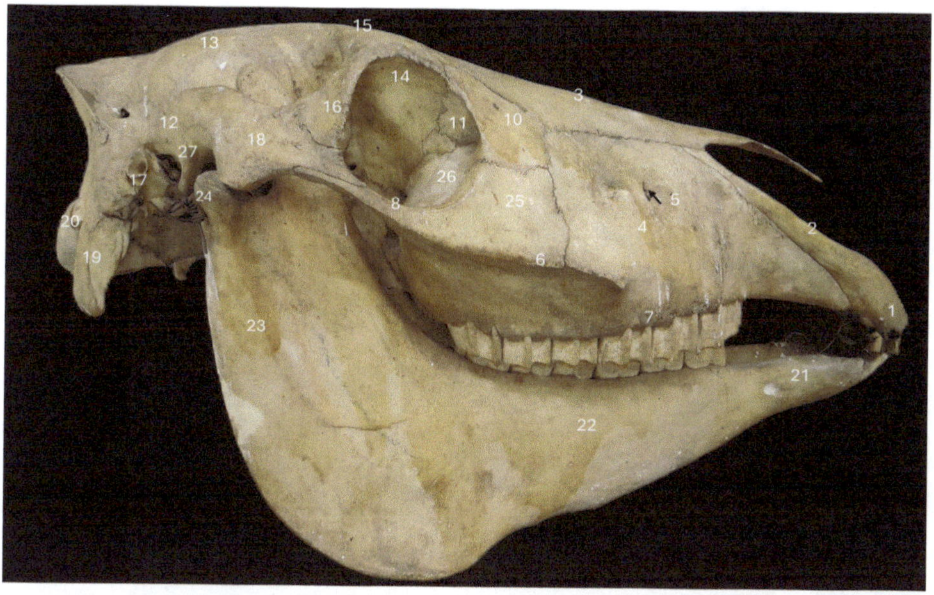

Fig. 2.11: Photograph showing the different cranial and facial bones of the skull of horse

1. Body of premaxilla
2. Nasal process of premaxilla (facial surface)
3. Nasal bone
4. Body of maxilla (facial surface)
5. Infra orbital foramen
6. Facial crest
7. Alveolar border of maxilla
8. Zygomatic process of malar
9. Orbital margin of lacrimal
10. Facial surface of lacrimal
11. Orbital surface of lacrimal
12. Temporal
13. Vertical part of parietal
14. Orbital surface of frontal
15. Facial surface of frontal
16. Supra orbital process
17. External acoustic meatus
18. Zygomatic process of temporal
19. Paramastoid process
20. Occipital condyle
21. Body of the mandible
22. Horizontal ramus of the mandible
23. Vertical ramus of the mandible
24. Condyle of the mandible
25. Facial surface of malar
26. Orbital surface of malar
27. Retroauricular process

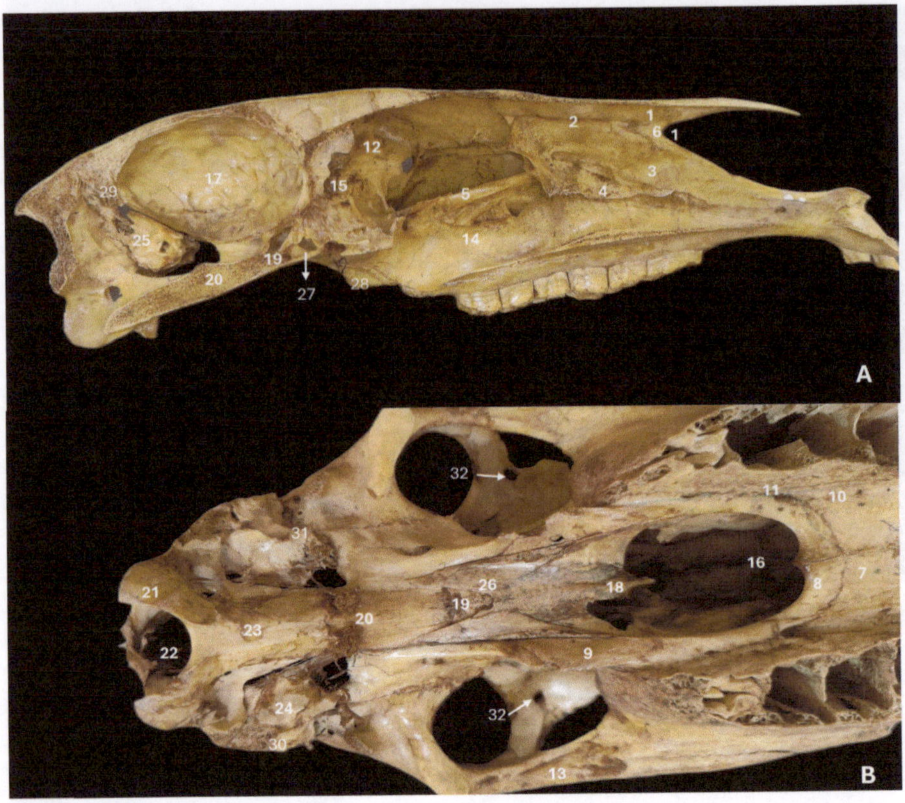

Fig.2.12: Photograph showing the Skull of Horse: A. Sagittal section, B. Ventral view

1. Nasal bone (nasal surface)
2. Crest for dorsal turbinate bone
3. Nasal surface of maxilla bone
4. Crest for ventral turbinate
5. Infra orbital canal
6. Naso-maxillary notch
7. Palatine surface of maxilla
8. Palatine surface of palatine bone
9. Perpendicular plate of palatine bone
10. Palatine groove
11. Cranial palatine foramen
12. Nasal surface of lacrimal bone
13. Temporal process of Zygomatic bone
14. Maxillary sinus
15. Ethmoturbinat
16. Nasal cavity
17. Cranial cavity
18. Vomer
19. Pre-sphenoid
20. Post sphenoid
21. Occipital condyle
22. Foramen magnum
23. Basioccipital
24. Tympanic bulla
25. Petrous temporal
26. Pterygoid
27. Sphenoidal sinus
28. Sphenoid process of palatine bone
29. Ossium tentorium
30. External acoustic meatus
31. Styloid process
32. Opening of supra orbital canal

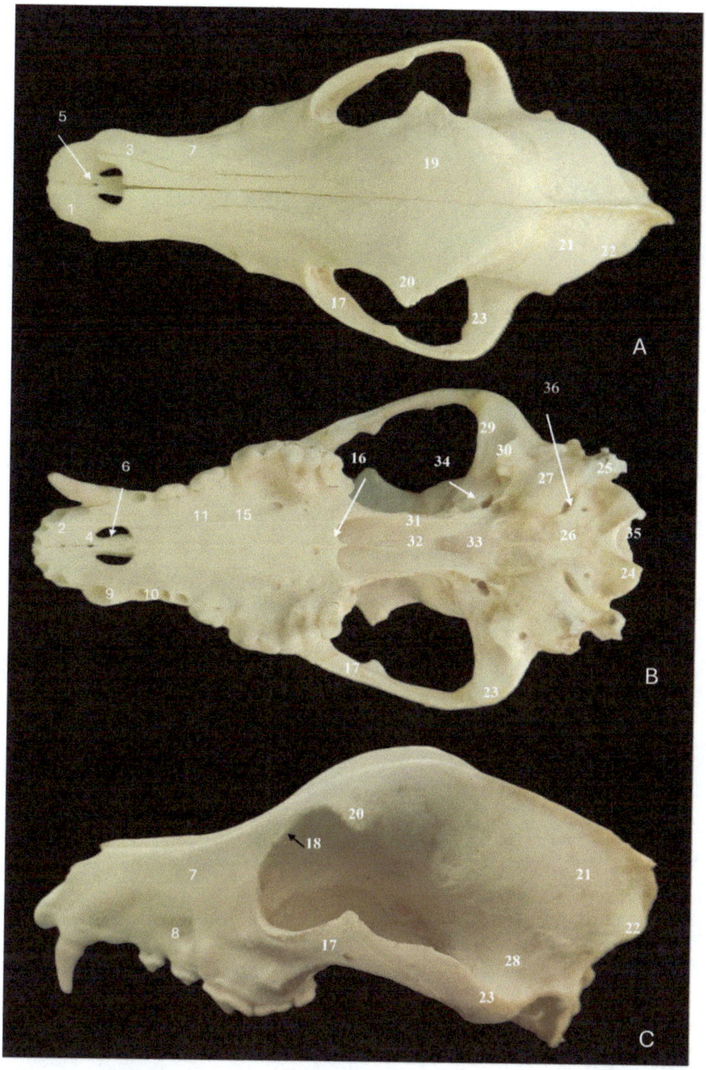

Fig. 2.13: Photograph showing cranial and facial parts of Dog Skull A. Dorsal view, B. Ventral view, C. Lateral view, 1. **Labial surface of body of premaxilla, 2. Palatine surface of body of premaxilla,** 3. Nasal process of premaxilla (facial surface), 4. Palatine process of premaxilla (palatine surface), 5. Incissive foramen, 6. Palatine fissure, 7. Body of maxilla (facial surface), 8. Infra orbital foramen, 9. Alveoli for canine, 10. Alveolar border of maxilla, 11. Palatine process of maxilla, 12. Cranial palatine foramen, 13. Horizontal part of palatine bone, 14. Perpendicular part of palatine bone, 15. Palatine groove, 16. Caudal nasal spine of palatine, 17. Temporal process of malar, 18. Orbital surface of lacrimal, 19. Frontal bone, 20. Supra orbital tubercle, 21. Parietal, 22. Nuchal crest, 23. Zygomatic process of temporal, 24. Occipital condyle, 25. Paramastoid process, 26. Basioccipital, 27. Tympanic bulla, 28. Temporal fossa, 29. Mandibular fossa, 30. Retroauricular process, 31. Pterygoid, 32. Pre sphenoid, 33. Post sphenoid, 34. Foramen oval, 35. Foramen magnum, 36. Jugular foramen

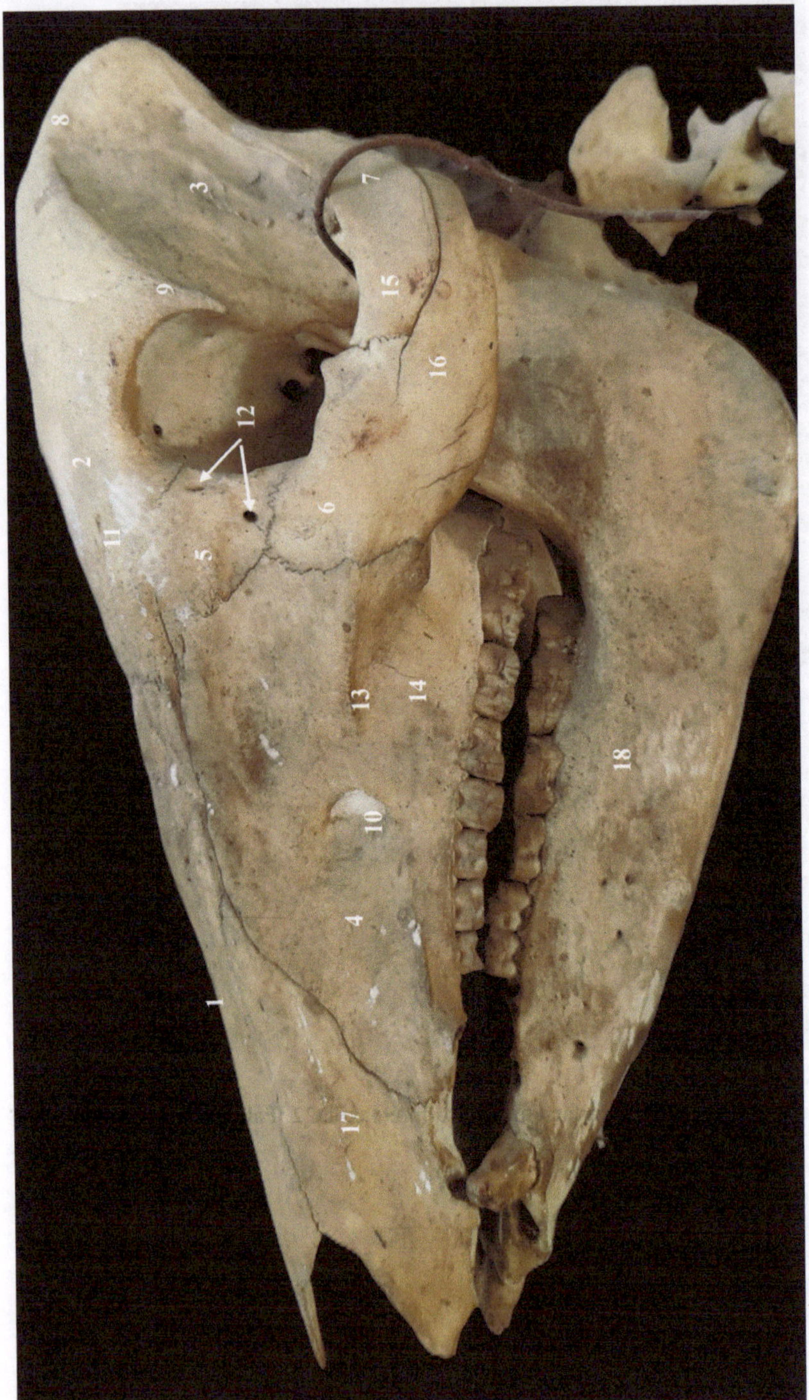

Fig. 2.14: Photograph showing the lateral view of skull of pig, 1. Nasal bone, 2. Frontal bone, 3. Parietal bone, 4. Maxilla bone, 5. Lacrimal bone, 6. Malar bone, 7. Temporal bone, 8. Nuchal crest, 9. Supra orbital process of frontal, 10. Infra orbital foramen, 11. Supra orbital foramen, 12. Lacrimal foramen, 13. Facial crest, 14. Muscularfossa, 15. Zygomatic process of temporal, 16. Temporal process of malar, 17. Premaxilla, 18. Mandible

B. Comparative Study of Facial Bones of Skull

Facial bones form the skeleton of the oral and nasal cavities and also support the pharynx, larynx and root of the tongue.

Comparative Study of Maxilla/ Superior Maxillary Bone of Ox

Maxilla is a paired and the largest bone of upper jaw and is situated on lateral aspect of the face and articulates with most of the facial bones (nasal, premaxilla, lachrymal, malar, and palatine). It has a body and a palatine process/ plate.

The body forms the lateral part of maxilla (Fig. 2.15). It has two surfaces, two borders and two extremities. **Facial/lateral surface** is concave infront and convex behind. It bears a rough facial tuberosity at the level of 3rd and 4th upper cheek tooth. A rough line, facial crest runs upward and backwards from facial tuberosity which joins the facial crest of malar bone. External opening of infra orbital canal i.e. infra orbital foramen which is often double is situated at the level of 1st cheek teeth. Nasal/medial/ internal surface forms a large part of nasal cavity. It is concave from above downwards. A ventral turbinate crest is situated about middle of this surface into which ventral turbinate bone is attached. Immediately below this crest a lacrimal groove runs which contain nasolacrimal duct. Parallel to crest a horizontal bony plate of palatine process is present which separates the nasal cavity from palatine sinus. Posterior part of this surface forms the part of maxillary sinus. Infraorbital canal forms the floor of maxillary sinus and continues as alveolar canal. The dorsal border is irregular. Its anterior most part is grooved to articulate with nasal process and premaxilla and further back articulate with nasal bone (sutura serrata). Posterior part of this border articulates with lacrimal and malar bones. The ventral/ alveolar border is very thick and having six alveoli for upper cheek teeth. The anterior one third part of this border infront of first premolar is concave, narrow and smooth and forms inter-dental space. A rough eminence, alveolar tuberosity is present behind last molar teeth which give origin to *buccinators and depressor labii inferioris muscle*. The posterior/caudal extremity presents maxillary tuberosity which contains posterior prolongation of maxillary sinus. Medial to it, there is a deep recess, maxillary hiatus, in which three foramen are lodged. Dorsal narrow fissure shaped foramen is maxillary foramen leads to infraorbital canal, middle foramen is sphenopalatine foramen opens into nasal cavity and ventral is caudal/posterior palatine foramen enter to palatine canal. Out of these three foramina spheno- palatine foramen is oval in shape and is largest. A small zygomatic process extends dorsolateral to maxillary tuber. Dorsomedially it joins with zygomatic process of temporal bone. A small pointed process, pterygoid process is projecting inward at the upper most part of the maxillary tuberosity (Fig. 2.15).

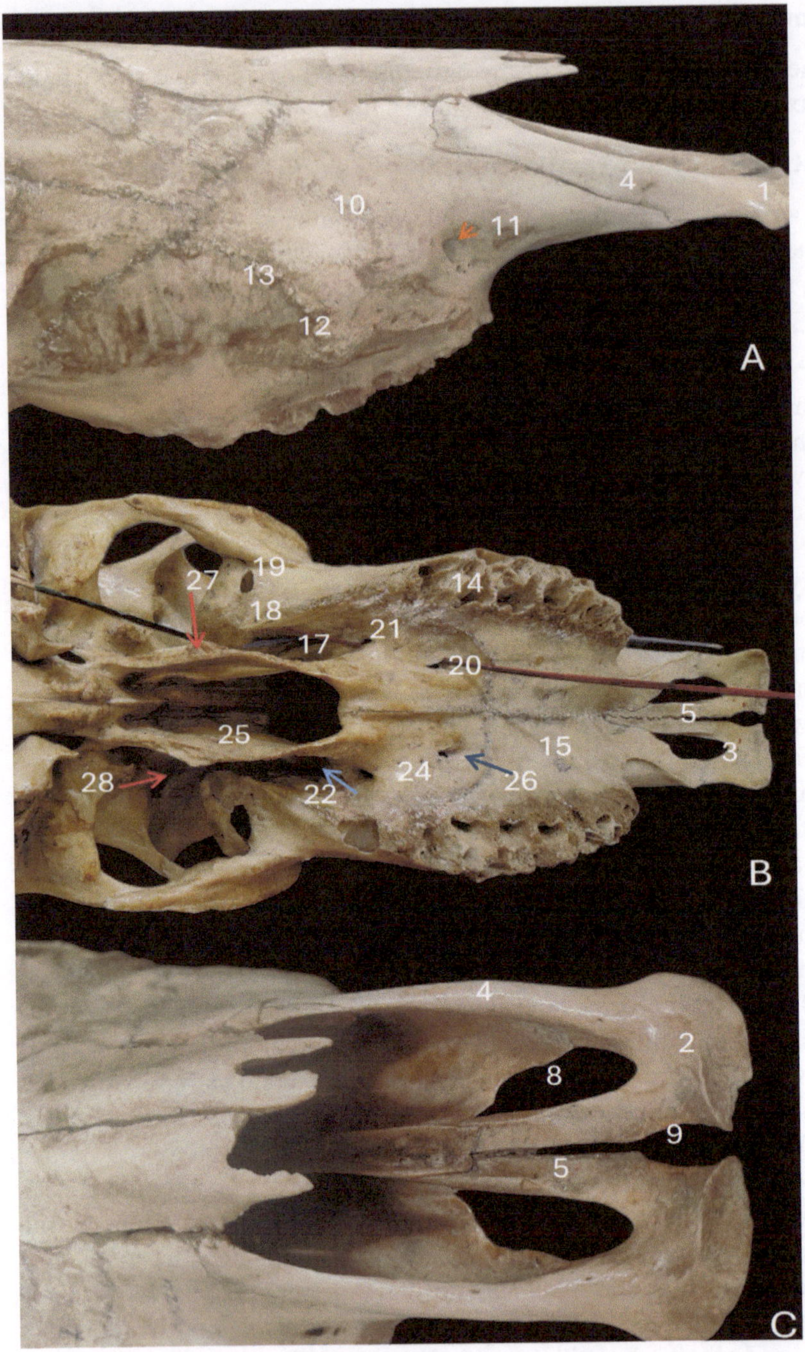

Fig. 2.15: Photograph showing the Premaxilla, Maxilla, Palatine and Pterygoid bones of Ox A. Lateral view, b. Ventral view and C. Dorsal view

Premaxilla

1. Body of premaxilla
2. Labial surface of body
3. lingual surface of body
4. Nasal process of premaxilla (facial surface)
5. Nasal process of premaxilla (nasal surface)
6. Palatine process of premaxilla (palatine surface)
7. Palatine process of premaxilla (nasal surface)
8. Naso palatine fissure
9. Incissive fissure

Maxilla

10. Body of maxilla (nasal surface)
11. Infra orbital foramen
12. Facial tuberosity
13. Facial crest
14. Alveoli for cheek tooth
15. Palatine process of maxilla
16. Alveolar border of maxilla
17. Maxillary hiatus
18. Maxillary tuberosity
19. Zygomatic process of maxilla
20. Cranial palatine foramen
21. Caudal palatine foramen
22. Maxillary foramen
23. Spheno-palatine foramen

Palatine

24. Horizontal part of palatine bone
25. Perpendicular part of palatine bone
26. Palatine groove

Pterygoid

27. Hamular process
28. Pterygo palatine fossa

The Palatine process forms a part of the roof/ hard palate of the mouth. The part is smooth and slightly concave from side to side (medial to lateral). It presents palatine groove, palatine canal and accessory palatine foramina. Palatine groove descends from the middle of this surface and continue posteriorly into palatine canal at the anterior palatine foramen. The nasal surface encloses a large palatine sinus by a thin horizontal bony plate. It has four borders. The lateral border unites with the body of maxilla while medial border articulates with its opposite fellow by denticulate suture and form medial palatine suture. On nasal surface this suture bear's nasal crest which along with its opposite fellow forms groove for Vomer bone. The caudal border articulates with palatine bone by transverse palatine suture. The cranial border is concave and forms lateral boundary of palatine fissure.

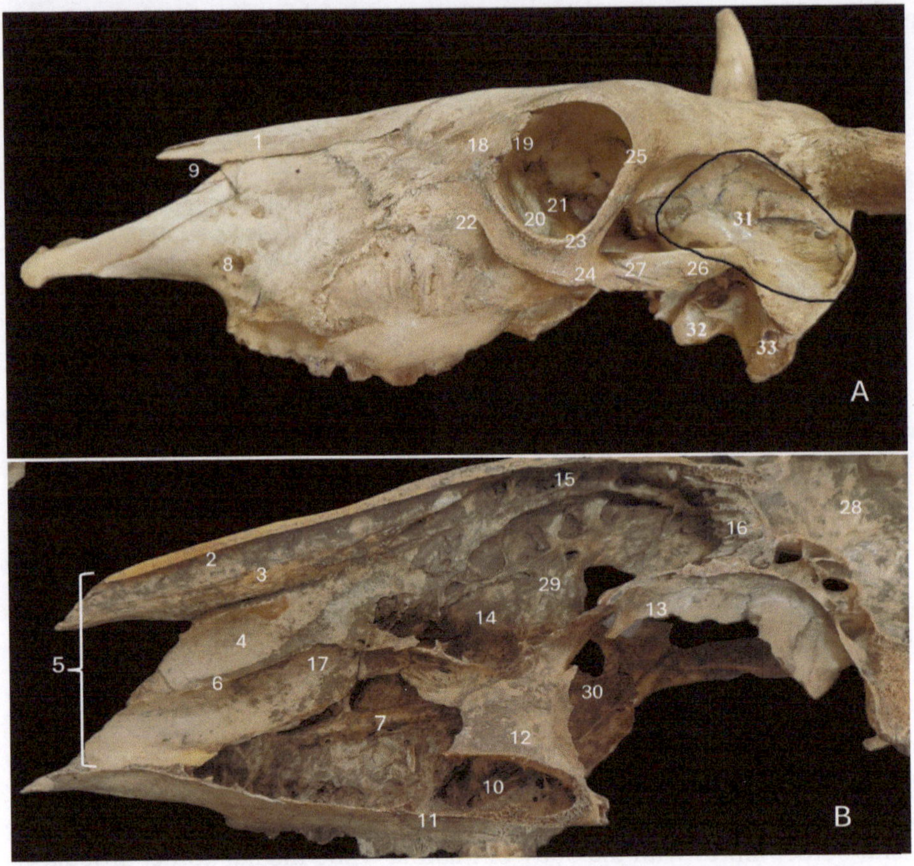

Fig. 2.16: Photograph showing the skull of Ox A. left lateral view, B. Cross section of skull

1. Nasal bone (facial surface)
2. Nasal bone (nasal surface)
3. Crest for dorsal turbinate bone
4. Nasal surface of maxilla bone
5. Nasal cavity
6. Crest for ventral turbinate
7. Infra orbital canal
8. Infra orbital foramen
9. Naso-maxillary notch
10. Palatine sinus (a. maxillary part, b. palatine part)
11. Palatine surface of palatine bone
12. Palatine plate
13. Perpendicular plate of palatine bone
14. Maxillary sinus
15. Frontal sinus
16. Ethmoturbinate
17. Groove for naso- lacrimal duct
18. Facial part of lacrimal bone
19. Orbital part of lacrimal bone
20. Orbital ring
21. Lacrimal bulla
22. Facial surface of zygomatic bone
23. Supra orbital process of zygomatic bone
24. Zygomatic process of zygomatic bone
25. Supra orbital process of fronatal bone
26. Zygomatic process of temporal bone
27. Zygomatic arch
28. Cranial cavity
29. Nasal surface of lacrimal bone
30. Nasal surface of zygomatic bone
31. Temporal fossa
32. Tympanic bulla
33. External Acoustic meatus

Table 2.8: Comparative table on Maxilla Bone of Ox, Horse, Dog and Pig

Characters	Ox	Horse	Dog	Pig
Relation	Incisive/premaxilla, nasal, frontal, lacrimal, malar, ventral nasal conchal bones	Same as in ox	Same as ox	Same as ox except it has also relation with Ethmoid bone
Facial crest	Present but less prominent than horse	More prominent than ox	Absent	Present but less prominent than horse
Facial tuberosity	Present dorsal to 3rd/ 4th upper cheek tooth	Absent	Absent	Absent
Infra orbital foramen	Located dorsal to the 1st cheek tooth	Two inches above the facial crest	Upper 3rd cheek tooth / premolar	Located just dorsal to the upper cheek teeth (premolars) and roughly at the level of the angle of the mouth (buccal commissure).
Alveolar canal	Not present	Not present	Not present	Located at the rostro-lateral part of the infra orbital foramen
Canine fossa	Not present	Not present	Not present	Present dorsal to facial crest and rostral to infra orbital foramen
Lacrimal groove	Present below the ventral turbinate crest	Shallow, Present above the ventral turbinate crest	Same as in ox	Same as in ox
Ventra turbinate crest	Long present above the lacrimal groove	Long present below lacrimal groove	Short and present rostrally	Short and extend rostro-laterally from a point dorsal to maxillary hiatus
Alveoli for canine	Absent	Present at the junction of maxilla and premaxilla	Present as in horse	Present as in horse
Interdental space	Concave , no alveoli for canine	Longer than ox, one alveolus for canine alveoli for canine	Smaller than ox and horse, one alveolus for canine	Shorter than ox and horse but longer than dog, one alveolus for canine
Articulation with	Articulate with premaxilla, nasal, lacrimal, and malar	Same as in ox	Articulates with premaxilla, frontal, malar and nasal only	Articulates with frontal, lacrimal, malar and nasal bones
Alveolar tuberosity	Present	Present	Absent	Absent

Characters	Ox	Horse	Dog	Pig
Maxillary tuberosity	Small laterally compressed	More prominent than ox	Smaller and less extensive structure compared to the horse and ox.	Relatively small, cone shaped and narrow in its middle portion.
Maxillary hiatus	Deep recess present medial to maxillary tuberosity	Same as in ox	Not formed instead of this small maxillary recess is present	Present, Incomplete in immature pig
Maxillary foramen	Formed in maxilla	Formed by maxilla and palatine bone	Formed by zygomatic process of maxilla and palatine bone	Large, oval shaped and located medial to zygomatic process
Sphenopalatine foramen	Formed in perpendicular part of palatine bone and completed by ethmoid and sphenoid bone	Formed by palatine bone	Formed by maxilla and palatine bone	Formed by maxilla, palatine and ethmoid bones
Caudal Palatine foramen	Formed by horizontal part of palatine bone	Formed by maxilla and palatine bone	Formed by maxilla and palatine bone	Formed by maxilla and palatine bone
Zygomatic process	Very Small and articulate with zygomatic process of malar bone	Large and articulates with temporal process of malar and zygomatic process temporal	Short and thin, overlapped by malar bone	Short , overlapped and articulate with malar bone
Pterygoid process	Small projection located behind the last alveolus.	Not present	Pointed projection	Not present
Frontal process	Absent	Absent	Present	Very small
Palatine process	Wider but shorter than horse	Narrow and longer than ox	Short and wider behind	Very long
Palatine groove	Shallow	Shallow	Distinct	Distinct
Palatine sinus	Present	Absent	Absent	Absent
Cranial palatine foramen	Present in palatine bone	Present in palatine bone	Located at or close to transvers palatine suture	Located at the palatine process of maxilla
Maxillary sinus	Small and undivided	Caudal part of the sinus is divided into medial and lateral parts by infra orbital canal	Participate	Same as in ox

Comparative Study of Premaxilla / Incisive Bone (*Os Incisivium*) of Ox

Premaxilla is a paired facial bone which forms the distal/ anterior part of upper jaw. Each bone has a body, palatine process and nasal process (Fig. 2.15). The body is thin and flattened and has no alveoli for incisor. Upper incisors and canines are absent in ox. The body has two surfaces viz. labial and palatine surface. The former one is convex, smooth and related with upper lip and later, palatine/ buccal surface is concave in both directions and related to dental pad or palate. The body has two borders. The cranial border is convex and separates the palatine and labial surface. Cranial border spring the palatine and nasal processes and forms the cranial boundary of palatine fissure. Medial border is concave and separated from its corresponding part of opposite bone by forming incisive fissure (Fig. 2.15 & 2.16).

Nasal process is larger and longer and projected upward and backward from the body. It forms the lateral wall of nasal cavity. It has two surfaces, two borders and two extremities. The facial/ external surface is smooth and convex and nasal surface is concave (Fig. 2.15). Its posterior two third part is applied against the anterior edge of body of maxilla. Anterior one third part is free, convex and smooth. Anterior extremity is thick and fused with body. Cranial extremity is thin and articulate with maxilla.

Palatine process is thin and flattened situated internally. It also has two surfaces, two borders and two extremities. Palatine surface is flat, smooth and form cranial part of hard palate. Nasal surface is grooved for vomer and septal cartilage. It forms cranial part of the floor of the nasal cavity. Media border is straight and lateral border is concave cranial which along with nasal process and maxilla bone forms palatine fissure/naso palatine cleft. Cranial extremity fused with body and posterior extremity is narrow, pointed and fits into interval between the vomer and palatine process of maxilla.

Table 2.9: Comparative table on Premaxilla Bone of Ox, Horse, Dog and Pig

Characters	Ox	Horse	Dog	Pig
Body	Thin and flattened	Thick	Compressed dorso-ventrally	Elongated and thicker dostrally and tapered caudally
Alveolus for Canine	Absent	Present (one)	Present (one)	Present (one)
Incisors	Absent	Present	Present	Present
Anterior border	Thin and convex	Contain alveoli	Contain alveoli	Contain alveoli
Incisive fissure	Wide	Absent	Absent	Narrow
Incisive foramen	Absent	Present	Very small	

Characters	Ox	Horse	Dog	Pig
Palatine process	Wide and extensive	Thinner than ox	short	Thin
Palatine fissure	Wide	Narrow	Oval , short and Wide	Present
Naso-maxillary notch	Present	Present	Semicircular	Large

Comparative Study of Palatine Bone of Ox

Palatine bone is a paired bone situated on either side of conchae and form caudal part of hard palate. Each articulates with opposite bone, maxilla, pterygoid, basisphenoid, vomer, presphenoid, ethmoid, frontal and sometime lacrimal bones also. It presents horizontal and perpendicular parts. Horizontal part form one fourth of the posterior part of hard palate. It has two surfaces viz. palatine and nasal surfaces and four borders (Fig. 2.15). The palatine surface is nearly flat and smooth and has anterior palatine foramen along with accessory palatine foramina. The cranial palatine foramen is the ventral / anterior opening of palatine canal. A groove, palatine groove runs anteriorly from this foramen. Nasal surface is slightly concave, smooth and hollow to form part of palatine sinus. This cavity continues below with that of palatine process of maxilla. The medial border meets its fellow at the median palatine suture. Its nasal part is in the form of rounded ridge. Cranial border is nearly straight in some or concavo-convex or entirely concave. It joins the palatine process of the maxilla bone by transverse palatine suture. The lateral border articulates with maxilla and serves to complete the alveolus for the last molar (Fig. 2.15). Cranial border is concave and narrow. V shaped plate spring form this border. Cranial palatine foramen is situated either or immediately behind the edge of this border.

The V shaped plate is an extensive, quadrilateral thin plate which forms the posterior part of the lateral wall of the nasal cavity and in part bound the posterior nares (Fig. 2.15). Its nasal surface is nearly flat, smooth and free except behind where it overlapped by pterygoid bone. The lateral surface is attached to small extent to pterygoid process behind and is free elsewhere. It forms the inner limit of pterygo-palatine fossa. The dorsal border presents a deep notch to form spheno-palatine foramen. The edge behind this foramen articulates with orbital wing of sphenoid bone. The ventral border is free at anterior part and remaining part is concealed by the overlapping pterygoid bone internally and pterygoid process externally.

Table 2.10: Comparative table on Palatine Bone of Ox, Horse, Dog and Pig

Characters	Ox	Horse	Dog	Pig
Part of hard palate formed	≥1/4	Caudal part	One third	One fifth
Location of cranial palatine foramen	1.5 cm from median palatine suture	On the transverse suture	On the transverse suture near lateral border	Towards lateral border (palate maxillary suture
Number of accessory palatine foramen	One	Same as ox	Variable in number	Variable in number
Perpendicular part	Extensive, quadrilateral thin plate	More extensive than horizontal part	Very extensive	Roughly traingular
Horizontal part	Excavated to form palatine sinus with palatine process of maxilla	Non excavated	Same as in horse	Same as in horse
Palatine sinus	Present	Perpendicular plate encloses the palatine sinus	Absent	Absent
Nasal crest	Rounded ridge	In the form of ridge	Incomplete	Complete, articulates with vomer
Caudal nasal spine	Absent	Absent	Present	Absent
Perpendicular plate	Quadrilateral	Thin, vertical, wing-like	Irregularly triangular	Triangular ,overlapped by maxilla forming a part of palatine canal
Spheno-palatine foramen	Formed between palatine, ethmoid and sphenoid	Formed in perpendicular plate of palatine	Formed in palatine bone	Formed between maxilla, palatine and ethmoid bones
Dorsal border of perpendicular plate	Articulate with presphenoid	Articulate with frontal	Articulate with frontal and sphenoid	same as in dog
Spheno-palatine sinus	Sphenoid sinus and palatine sinus are separate	Present	Absent	Present

Characters	Ox	Horse	Dog	Pig
Pyramidal /pterygoid process	Absent	projects posteriorly and laterally from the junction of the horizontal and perpendicular plates of the palatine bone.	absent	Present
Maxillary foramen	Situated in maxillary bone	Situated in maxillary bone	situated between zygomatic process of maxilla and perpendicular part of palatine bone.	Situated in maxillary bone

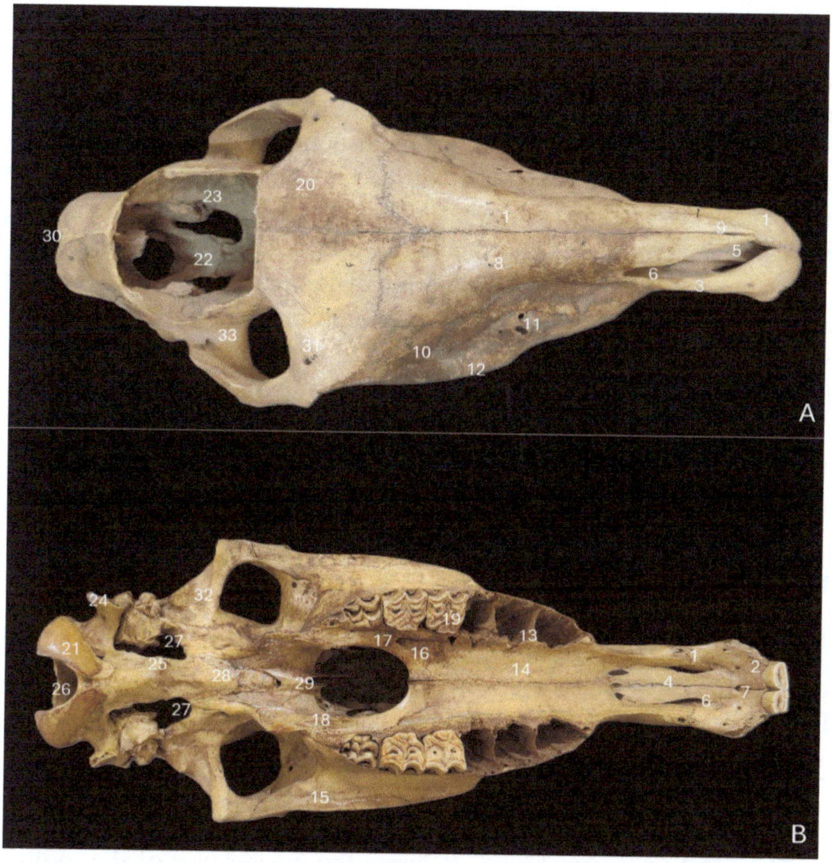

Fig. 2.17: Photograph Showing the A. Dorsal B. Ventral view of skull of Horse

1. Labial surface of body of premaxilla
2. Palatine surface of body of premaxilla
3. Nasal process of premaxilla (facial surface)
4. Palatine process of premaxilla (palatine surface)
5. Palatine process of premaxilla (nasal surface)
6. Naso palatine fissure
7. Incisive foramen
8. Nasal bone
9. Nasal peak
10. Body of maxilla (facial surface)
11. Infra orbital foramen
12. Facial crest
13. Alveoli for cheektooth
14. Palatine process of maxilla
15. Zygomatic process of maxilla
16. Cranial palatine foramen
17. Horizontal part of palatine bone
18. Perpendicular part of palatine bone
19. Palatine groove
20. Frontal bone
21. Occipital condyles
22. Cranial surface of basioccipital
23. Cranial cavity
24. Paramastoid process
25. Basilar tubercle
26. Foramen magnum
27. Jugular foramen
28. Sphenoid bone
29. Vomer bone
30. Nuchal crest
31. Supra orbital foramen
32. Zygomatic process of temporal
33. Temporal fossa

Comparative Study of Pterygoid Bone of Ox

Pterygoid is a paired narrow, thin bent like plate of bone situated obliquely downward and forward on either side of conchae. It articulates with palatine and pterygoid process of sphenoid bone. Each bone has two surfaces, two borders and two extremities. The medial surface is free, smooth and form the lateral wall of nasal cavity. Most of the parts of the lateral surface are concealed by palatine and pterygoid process of sphenoid bones. Its caudal most part is free and can be seen in pterygo-palatien fossa. The dorsal border is irregular and attached with palatine bone, whereas, ventral border is free. The cranial extremity is free and its ventral end is projected in the form of hook known as pterygoid hamulus/ hamular process. Tensor palati muscle reflects around this process. The caudal extremity is wide and wedges in between the pterygoid process and body of sphenoid bone.

Horse: It is shorter and narrower twisted bone placed obliquely. Hamulus is grooved externally and forms a pulley. It articulates with palatine, vomer and sphenoid bone.

Dog: It is wide and quadrilateral short bone. Hamulus is variable.

Pig: It is vertical in direction, narrow in middle and wider at the end. It forms the middle wall of pterygoid fossa. Hamulus is blunt.

Comparative Study of Nasal Bone of Ox

Nasal bone is a paired bone situated on the anterior aspect of the face between frontal, lacrimal and maxillary bones. It forms the greater part of the roof of the nasal cavity. It presents two surfaces, two borders and two extremities. The facial surface is convex from side to side and smooth, nearly straight from above downward. The nasal surface is concave towards medial border in the form of a deep longitudinal groove and forms dorsal nasal meatus. Immediately below the groove a turbinate crest is present for attachment of dorsal turbinate bone. Anterior to the crest the surface is slightly convex or flat. Posterior-lateral part of the crest participates in the formation of frontal sinus. The medial border is straight and plain to form sutura hormonia with opposite fellow. The lateral border is irregular and articulate with maxilla, lacrimal and frontal bones. Posterior extremity is pointed to form alar projection which received into the notch formed by united anterior extremity of frontal bones. Anterior extremity is divided into two parts by a deep notch. The inner processe of the two bones lie close together and thus gives the nasal peak a trifid appearance (Fig. 2.16 and 2.18).

Table 2.11: Comparative table on the Nasal Bone of Ox, Horse, Dog and Pig

Characters	**Ox**	**Horse**	**Dog**	**Pig**
Bone	Long and wide	Longer and wider than ox	Long wider infront than behind	Long wider infront than behind
Facial surface	Concave	Convex	Concave	Convex
Cranial extremity	Divided	undivided	Undivided, formed semicircular nasal notch	Undivided
Nasal peak	Trifid	Not trifid	Absent	Absent
Processus septalis	Not present	Not present	Present, it forms the part of nasal septum	Not present
Relation with frontal bone	Alar projection received into the notch formed by united anterior extremity of frontal bones	United upper extremity of two nasal bones forms a notch into which frontal bone is received	Sane as in ox	Same as in ox
Internal ethymoidal crest	Absent	Absent	Present in the form of longitudinal ridge on the concave internal surface	Present
Supra orbital sulcus	Absent	Absent	Absent	Located on facial surface in an oblique manner caudolaterally

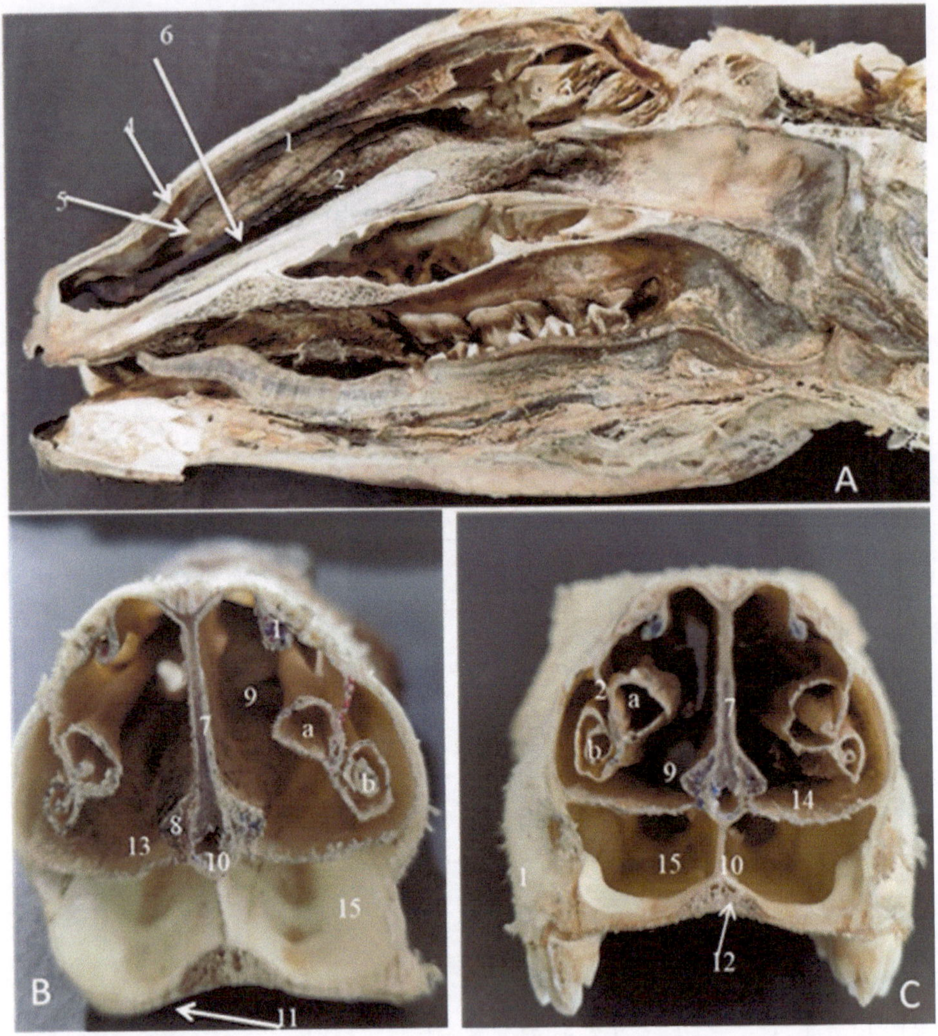

Fig. 2.18: Photograph showing the cross section of head of ox (A), anterior part of nasal cavity (B) and middle part of nasal cavity (C)

1. Dorsal turbinate
2. Ventra turbinate
   a. Dorsal part
   b. Ventral part
3. Ethmoturbinate
4. Dorsal meatus
5. Middle meatus
6. Ventral meatus
7. Nasal septum
8. Vomeronasal organ
9. Nasal cavity
10. Vomer
11. Palatine surface of maxilla
12. Palatine surface of palatine bone
13. Nasal surface of maxilla
14. Nasal surface of palatine
15. Paltine sinus

Comparative Study of Turbinate Bone of Ox

Turbinates are paired delicate, scroll like bones situated ventrally in the nasal cavity and attached to its lateral walls. These are fragile and cribriform in nature. In fresh state, they are covered by mucous membrane. These are also called as nasal choncha (Fig. 2.18).

Dorsal nasal turbinate bone is somewhat cylindrical, less cribriform and fragile and widest in the middle. It is attached to turbinate crest of nasal bone and nasal plate of frontal bone. Dorsal nasal meatus is situated between dorsal nasal turbinate and roof of nasal cavity. The middle nasal meatus is situated between ventral surface and ventral turbinate bone. It is separated from the nasal septum by a narrow interval the common nasal meatus. Its anterior part is rolled like scroll one and half times while posterior extremity is not rolled, small and joins the cribiform plate and lateral mass of ethmoid bone.

Ventral turbinate is consisted similar to dorsal one with an anterior coiled and posterior uncoiled portion. It is shorter but broader and attached to the maxilla by basal lamella about an inch wide which split into two plates rolled in opposite directions and enclosed two separate cavities, subdivided by several septa. The dorsal one opens into middle nasal meatus and ventral one into ventral nasal meatus. Ventral nasal meatus which is much larger than the other meatuses is situated between floor of nasal cavity and ventral turbinate bone.

Horse: Dorsal turbinate is more cribriform, fragile, longer and larger than ox. Ventral turbinate forms only single scroll. The caudal part of ventral nasal choncha contains large bulla dorso laterally.

Dog: The turbinates are richly convoluted (secondary and tertiary laminae). They do not communicate with the sinuses of the skull.

Pig: scroll of dorsal turbinate is present throughout the length. Rostrally it is open and caudally ends in a closed spiral enclosing a recess of the nasal cavity.

Comparative Study of Lacrimal Bone of Ox

Lacrimal is a paired bone situated at rostral part of orbit and extend upto caudal border of maxilla. Its circumference is very irregular, thin and scaly which is formed by the articulation with frontal and nasal bones, ventrally with malar and maxilla bones, Rostrally with maxilla and caudally with frontal bone. It presents three surfaces *viz.* orbital, facial and nasal/ internal (Fig. 2.15 and 2.16). A distinct orbital margin divides the facial and orbital surface. The orbital part is smooth and concave and forms anterior wall of the orbit. A shallow lacrimal fossa is present just behind the orbital margin which lodges the lacrimal sac. Entrance of lacrimal canal located at the bottom of the fossa into which naso lacrimal duct runs. The orbital part bears ventrally a large, very thin walled protuberance, Lacrimal bulla, which bulges backward into the lower part of orbit and contains extension of maxillary sinus. The facial part is extensive and concave in its length. Nasal/ Internal surface is irregularly concave. Lacrimal canal divides this surface into two unequal halves. Both the parts participate in the formation of maxillary sinus.

Table 2.15: Comparative table on Lacrimal Bone of Ox, Horse, Dog and Pig

Characters	Ox	Horse	Dog	Pig
Size	Large	Smaller than ox	Very small	Small
Articulation	Frontal, malar, maxilla and nasal bones	Same as in ox	Frontal, malar, maxilla and palatine bones	Frontal, malar, maxilla bones
Orbital surface	Triangular	Funnel shaped	Small and triangular	Small
Lacrimal bulla	Present	Absent	Absent	Present
Facial part	Large	Less extensive than ox	Very much negligible	Less extensive than ox
Lacrimal foramen	Single	Single	Single	Paired
Facial crest	Absent	Absent	Absent	Continuous ventrally with the facial crest of malar and maxilla bone
Lacrimal tubercle	Absent	Present one inch below the orbital ring	Absent	Absent

Comparative Study of Malar or Zygomatic Bone of Ox

Malar is a paired triangular bone situated between lacrimal and maxilla. It presents orbital, nasal and facial (lateral) surfaces and cranial, dorsal and ventral borders (Fig. 2.16). Facial surface is extensive and bears facial crest just below the orbital margin which is continue on the maxilla. The surface is concave dorso- ventrally below the facial crest. The orbital surface is smaller, concave and smooth. It forms the floor and outer wall of the orbit. The nasal/ internal surface is free, concave and forms part of the maxillary sinus. The dorsal border of facial part is denticulated and articulate with lacrimal. The orbital part of dorsal border is concave and forms part of orbital ring. Ventral and cranial borders articulate with maxilla. The internal border of orbital surface is concave and free. It is separated from lateral border of lacrimal bulla by an extremely narrow interval.

Table 2.13: Comparative table on Zygomatic Bone of Ox, Horse, Dog and Pig

Particulars	Ox	Horse	Dog	Pig
Facial, orbital and nasal parts	All three parts are Present	As in ox	As in ox but Facial part is very small	As in ox, same as in ox
Frontal process	runs upward and backward and joins the supra orbital process of frontal bone	Absent	In form of tubercle	Small and sharp
Supra orbital margin	Formed by frontal process of malar and zygomatic processof frontal bone	Formed by zygomatic process of frontal and zygomatic process temporal bone	Not present instead of this orbital ligament connects the frontal process of zygomatic and zygomatic process of frontal to complete the orbital ring	Same as in dog
Temporal process	Runs backward and overlapped by zygomatic process of temporal bone to form zygomatic arch	Same as in ox	Large, long and strongly curved	Large and pointed
Maxillary process	Absent	Absent	Directed ventrally	Absent
Lacrimal process	Absent	Absent	Projected dorsally and fitted between lacrimal and maxilla bone	absent
Facial crest	"S" shaped, continue rostrally with facial crest of maxilla	Very prominent, rostrally continue with facial crest of maxilla and caudally with zygomatic process	Less distinct	Less distinct
Muscular fossa	Not present	Not present	Not present	Present on facial surface along with maxilla and lacrimal bone form muscular fossa

The posterior extremity is formed by zygomatic process which divides into two branches. The frontal branch/process runs upward and backward and joins the supra orbital process of frontal bone. The temporal branch/ process continued backward and overlapped by zygomatic process of temporal bone to form zygomatic arch.

Comparative Study of Vomer Bone of Ox

Vomer is the unpaired median bone which forms the ventral part of nasal septum and extends above downward from body of presphenoid to the groove of palatine process of maxilla. It is in the form of very wide thin plate which has two surfaces, two borders and two extremities. The two lateral surfaces are wider in upper two third part and gradually reduced towards the anterior/ lower third part. The surfaces are flattened or slightly convex and in fresh state covered with nasal mucous membrane. The cranial/ dorsal border has a wide deep groove, sulcus vomeris. The groove receives ventral part of the perpendicular plate of ethmoid bone caudally and nasal septum cranially. The caudal/ventral border is slender in its lower third part lies in a groove formed by nasal crest of palatine process of maxilla. Above this it is free and caudally forms a slit like space through which the posterior nares communicate with each other. The lower / cranial extremity articulates in a groove formed by the palatine process of premaxilla. The upper/ caudal extremity is applied against the body of preshenoid. It also articulates with the upper extremity of pterygoid bone (Fig. 2.15 and 2.18).

Table 2.14: Comparative table on Vomer Bone of Ox, Horse, Dog and Pig

Characters	Ox	Horse	Dog	Pig
Sulcus vomeris	Deep	Shallower than ox	Very deep compared to horse	Deep
Lateral wing	Absent	Present ,articulates with palatine, pterygoid and presphenoid bones	Absent meet only with palatine bone and helps in the formation of lamina transversalis	Present and articulates with palatine and body of presphenoid
Nasal fossa	Caudal border is free thus the caudal nares are interconnected	Completely divided by caudal border of vomer	Same as in ox	Same as in horse

Comparative Study of Mandible or Lower Jaw Bone of Ox

Mandible is the largest unpaired bone of face though it consist of two halves at birth. It carries the lower teeth and articulate by its condyles with squamous temporal on either side. It consists of body and two rami.

Body: Its thick cranial part formed by union of ventral end of two halves meet at symphysis mandible which bears the alveoli for incisor teeth. It presents three surfaces and two borders. Lingual surface is smooth and concave, labial/ mental surface is convex and extensive and symphyseal surface is rough, irregular and marked by projections and corresponding depressions. Alveolar border is convex which separates the lingual and labial surfaces. It presents eight alveoli for lower incisor teeth. The interalveolar border is long, curved, thin and sharp. Alveoli for canine is absent in this border.

Rami: Right and left rami are symmetrical, flattened from side to side and extend backward and upward. These diverge to enclose the "V' shaped mandibular space. Each ramus consists of a horizontal part which bears lower cheek teeth and bend upward at the angle to form vertical part. Vertical part is expanded and furnishes attachment for masseter muscle. Each ramus presents two surfaces, two borders and two extremities. At the junction of lateral surface of ramus and body there is an elliptical fossa which contains mental foramen at the caudal end of the fossa and is the external opening of mandibular canal. In the middle of medial surface of vertical ramus, there is mandibular foramen which is posterior opening of mandibular canal. A groove for lingual nerve runs downward and forward from the mandibular foramen. The alveolar border of horizontal part is straight which presents six alveoli for lower cheek teeth. Interdental space is situated between corner alveolus of incisor and first alveolus of premolar. The ventral border is convex, thick and rounded in horizontal part. The vertical part is convex and thin below and thick above. The lateral surface is widest at the angle and narrow at either extremity, smooth and convex in molar part and slightly concave in vertical part. Medial surface is smooth and slightly convex in horizontal part. Mylohyoid line is present on this surface close to alveolar border for attachment of mylohyoid muscle. On the vertical part medial surface is concave. It has a slight rough elevation at its lower end above the angle for medial pterygoid muscle.

Caudal extremity comprises the coronoid process infront and condyle behind separated by mandibular notch. The coronoid process is thin transversely and projects upward, backward and outward about two inches above the level of condyle. Condyles are elongated transversely and articulate with condyles on the squamous part of the temporal bone.

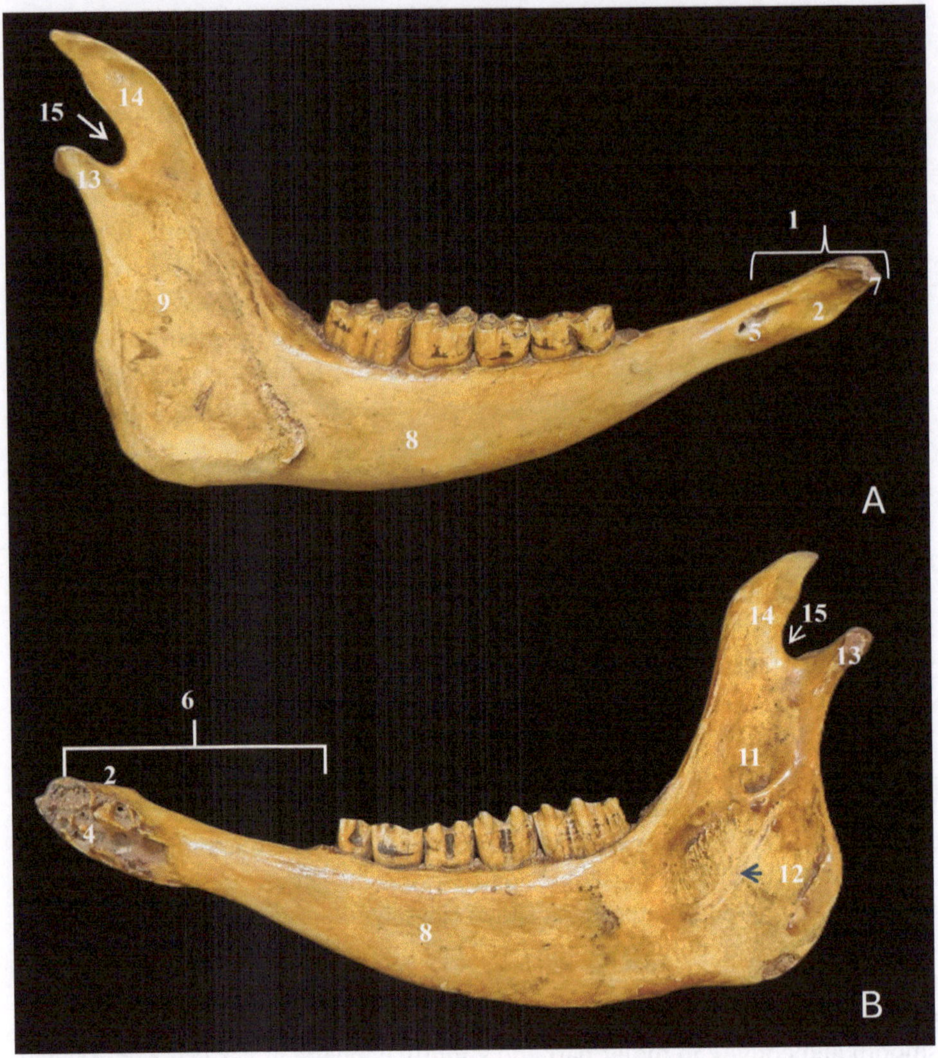

Fig. 2.19: Photograph Showing the Mandible of Ox A. lateral view, B. Medial view

1. Body of mandible
2. Lingual surface
3. Labial surface
4. Symphysial surface
5. Mental foramen
6. Interalveolar border
7. Alveolar border
8. Horizontal ramus
9. Vertical ramus
10. Angle of mandible
11. Mandibular foramen
12. Lingual groove
13. Condyle
14. Coronoid process
15. Mandibular notch

Table 2.15: Comparative table on Mandible Bone of Ox, Horse, Dog and Pig

Characters	Ox	Horse	Dog	Pig
Symphyseal surface	Present, do not fuse completely	Absent, fused completely	Same s in ox	Same as in horse
Alveoli	Eight for incisors and six alveoli for lower cheek teeth	Six incisors, two canines , six alveoli for lower cheek teeth	Six i alveoli ncisors two canines Seven for lower cheek teeth	Six incisors two canines Seven for lower cheek teeth
Condyle	Transversely elongated	Convex in both directions	Same as in ox	Same as in ox
Coronoid process	Strongly bent backward and outward, Wider below and tapering above	Thinner than ox and is rounded at free end.	Extensive plate like, as broad as ramus, free end is rounded	Small and thin
Masseteric fossa	Absent	Absent	Present ,located on the lateral surface of vertical ramus and extended upto coronoid process	Present
Mental foramen (mental/ lingual surface surface)	Absent	Absent	Absent	One pair
Mental foramen (lateral/ labial surface)	Single	Single	Two to three in number	Upto five openings
Angle of jaw	Present in between horizontal and vertical ramus	Same as in horse	In the form of angular process	Same as in ox but thick and smoothly curved
Diastemal/ inter-alveolar space	Long	Smaller than ox but longer than dog and pig	Very Small or absent	Small
Mandibular notch	Wide and deep	Same as in ox	Same as in ox	Wide
Lingual groove	Present	Absent	Absent	Absent
Angular process	Absent	Absent	Present	Absent

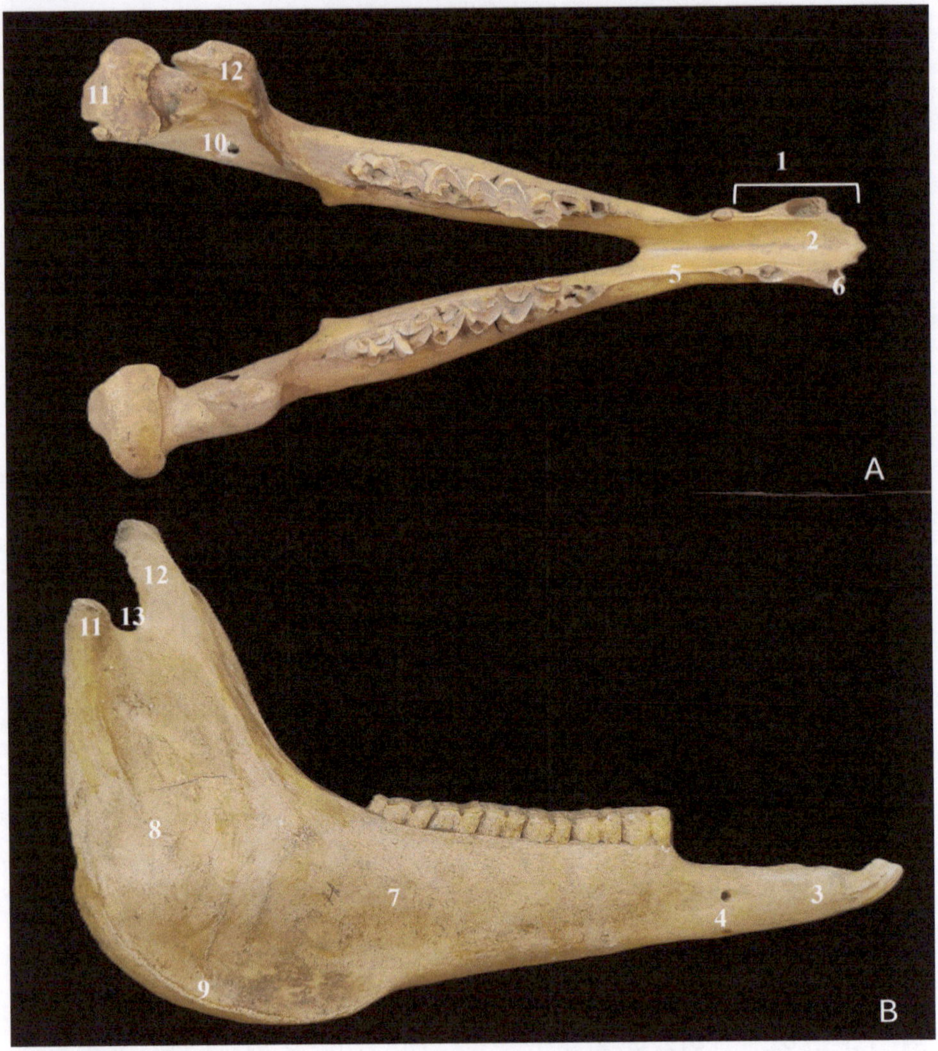

Fig. 2.20: Photograph Showing the Mandible of Horse A. Dorsal view, B. Lateral view

1. Body of mandible
2. Lingual surface
3. Labial surface
4. Mental foramen
5. Interalveolar border
6. Alveolar border
7. Horizontal ramus
8. Vertical ramus
9. Angle of mandible
10. Mandibularforamen
11. Condyle
12. Coronoid process
13. Mandibular notch

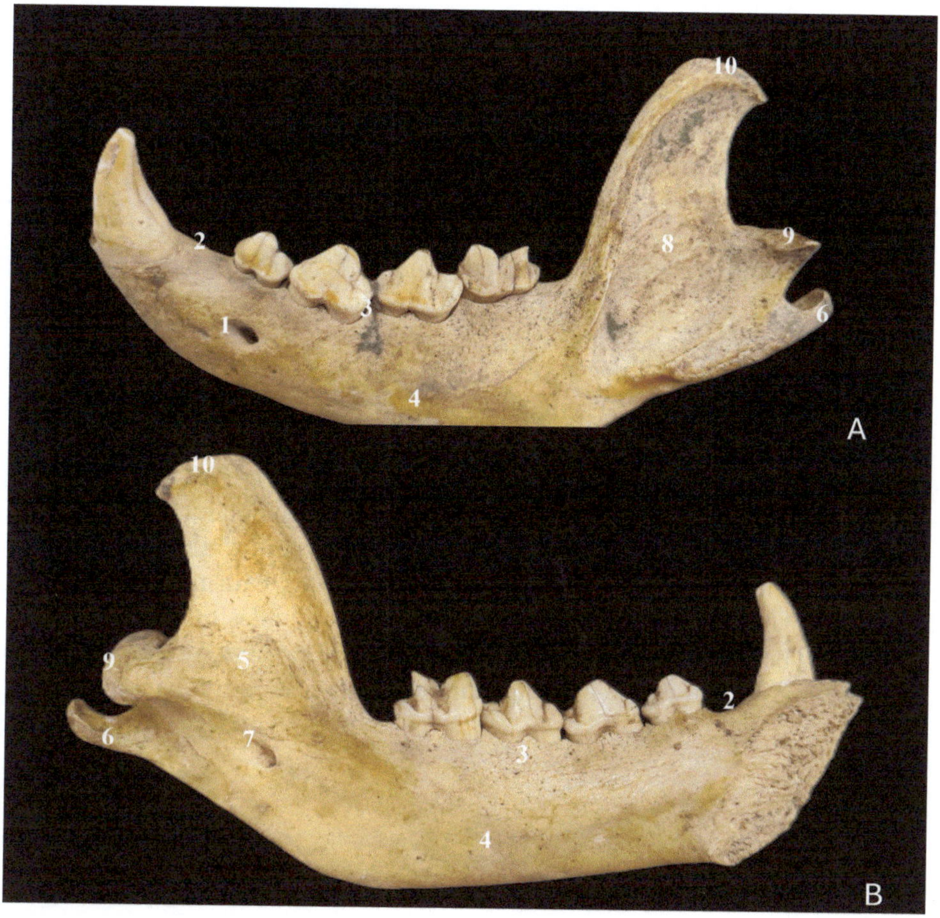

Fig. 2.21: Photograph showing the mandible of Dog A. lateral surface, B. Medial surface

1. Mental foramen
2. Interalveolar border
3. Alveolar border
4. Horizontal ramus
5. Vertical ramus
6. Angular process
7. Mandibularforamen
8. Masseteric fossa
9. Condyle
10. Coronoid process

Comparative Study of the Hyoid Bone of Ox

Hyoid bone is a single unpaired cartilaginous structure situated between rami of mandible and attached to the hyoid process of petrous temporal bone by a rod of cartilage, suspended obliquely downward and forward. It gives support to the root of tongue, pharynx and larynx. It is composed of body, lingual process and four pair of cornua viz. thyroid cornu (thyrohyoideus), Small cornu (keratohyoideus), middle cornu (epihyoideus) and great cornu (stylohyoid). The body or basihyoid is dorso ventrally compressed short transverse bar like structure. Its dorsal surface is convex and slopes downward and backward. Ventral surface is slightly concave and rough. *Lingual process* attached on the rostral border of the body. The lingual process is short and tuberous which projects rostrally and embedded in the root of the tongue. It gives attachment to *mylohyoid, geniohyoid and hyoglossus* muscles. Paired thyroid cornu is located on either side of the caudal border and extends upward and backward. It is flattened from side to side, wider infront and narrow behind. Its caudal end connects with the rostral cornu of thyroid cartilage of larynx by prolonged bar of cartilage. Paired small cornu/ keratohyoid/ ceratohyoid are short rods, flattened from side to side and directed upward and forward. Its ventral end articulates with the body by a facet and dorsal end connect with the middle cornu by a short piece of cartilage. Epihyoid/ middle cornua are almost as large as small cornua narrow below and wider above. Its dorsal end connects with large cornu by cartilage. Stylohyoid/ great cornua are largest part of the bone which is directed upward and backward. Its medial surface is convex while lateral surface is concave. The dorsal border is thin while ventral border is thicker than dorsal one. The ventral end connects with middle cornu by cartilage. The proximal extremity of styloid cornu divides into two branches by angle. The upper end articulates with the hyoid process of temporal bone by a piece of cartilage. The muscular angle is rough for muscular attachment.

Table 2.16: Comparative table on Hyoid Bone of Ox, Horse, Dog and Pig

Characters	Ox	Horse	Dog	Pig
Body	Short and flattened	Shorter than ox	Slightly curved transverse rod	Broad rostro-caudally and short transversely
Lingual process	Short	Longer, fused with body and has blunt pointed free end	Absent	Very short/ absent
Thyroid cornu	Fused with basihyoid	Same as in ox	Not fused but permanently connected with body by cartilage and form movable joint	Wide and form synchondrosis
Small cornu	Small rounded	Short rod	Short and prismatic	Short and wide
Middle cornu	Small rounded	Small wedge shaped	Longer than great cornu	Slightly longer than small cornu
Great cornu	Laterally flattened and cylindrical, About 6 cm long	Longer and thinner than ox 18-20 cm long	Bent outward and twisted, distal part remains cartilaginous	Slender rod shaped, distal part remains cartilaginous
Temporal bone articulation	Styloid process of tympanic part of temporal bone	Same as in ox	Mastoid process of the petrous temporal bone	Nuchal process of squamous temporal bone

B. Comparative Study of Bones of Vertebral Column

A series of vertebrae articulate together to form a long column of axial skeleton in the body known as vertebral column or spine. Each individual bone of the vertebral column is termed as vertebra (plural = vertebrae) which is defined as "an unpaired, irregular bone lying dorsally along the middle of the animal body. Anteriorly it joins with the skull whereas caudally it is prolonged beyond the trunk to form the skeleton of the tail. On the basis of position and shape these are classified into five regions which consist of variable number of vertebrae in different animals.

Name of vertebrae	Region form	Cattle	Horse	Dog	Pig	Fowl
Cervical (C)	Neck	7	7	7	7	14
Thoracic (T)	Back	13	18	13	14-15	7
Lumbar (L)	Loin	6	6	7	6-7	L + S =14
Sacral (S)	Croup	5	5	3	4	
Coccygeal (Cy)	Tail	18-20	15-21	20-23	20-23	6-7
Total number		49-51	51-57	50-53	51-56	41-42

Different Parts of A Vertebrae

All vertebra has a common basic shape which shows adaptation in the various region according to the local function. For description a vertebra is composed of a body, arch and different processes viz. dorsal spinous, ventral spinous, transverse, articular, mamillary and accessary process.

Body/ Centrum: It is the more or less cylindrical mass on which the other parts are constructed. It has two extremities and two surfaces. The cranial extremity is convex and caudal extremity is concave (Fig. 2.34). Both the extremities of the body are covered by a plate of hyaline cartilage which form the unossified part of the epiphysis of the body. The body of the adjacent vertebrae are attached with each other by intervertebral fibro-cartilages. The dorsal surface is flattened and forms the floor of the vertebral canal, while the ventral surface is rounded laterally and related with the various muscles and viscera. In the thoracic region the body presents two pairs of facets at the extremities for articulation with parts of the heads of two pairs of ribs (Fig. 2.31).

Vertebral / neural Arch: The arch form the lateral walls and roof of the vertebral canal. It consists of two lateral halves which is constructed on the dorsal aspect of the body. Each half is formed by pedicle and lamina (Fig. 2.24). The pedicle forms the lateral part of the arch. It has a cut infront and behind to form the cranial and caudal vertebral notches. The vertebral notches of two adjacent vertebrae join to form inter vertebral foramina for the passage of the spinal nerves and vessels. In some vertebrae, instead of notches complete foramina are present known as lateral vertebral foramina. The laminae are plate like structure uniting with each other medially at the root of the spinous process and complete the arch dorsally.

Processes: The vertebrae comprise of different processes. The form and size such as length and height and direction of these processes may be varied in different regions. Some process may be present or absent in some vertebrae. Following processes are generally present in the vertebra

1. **The Dorsal Spinous Process/ Supraspinous process:** It is single process which projects dorsally from the middle of the arch. It varies greatly in form, size and direction in different vertebrae. In furnishes attachment to muscles and ligaments (Fig. 2.24 and 2.31).
2. **The Ventral Spine:** It is single process placed in the middle of the ventral surface of the body of many vertebrae. This is not well developed in some vertebrae (Fig. 2.24).
3. **The Articular or Oblique Processes:** Two cranial and two caudal articular processes project from the borders of the arch (Fig. 2.24). They have articular facet to articulate with adjacent vertebrae to form inter-central articulation. The outer surface is roughened for muscular and ligamentous attachment.
4. **The Transverse Processes:** A pair of horizontal bony plates originated from the junction of the arch and body / neural arch and project laterally. In the thoracic region each has a facet on its ventral aspect, for articulation with a tubercle or a rib. They also give attachment to muscles and ligaments. Their size may vary according to the region of the vertebrae (Fig. 2.37).
5. **The Mamillary Process:** It is small, tubercle, like projection found in most animals. It is placed in between the cranial articular process and the transverse process or on the former or later, e.g., in the thoracic vertebrae of the ox (except at the caudal end of the series), on the caudal thoracic and cranial lumbar vertebrae in horse (Fig. 2.31).
6. **The Accessory Process:** It is situated in between the transverse process and the caudal articular process. e.g. in lumbar vertebrae of dog(Fig. 2.39).

Types of vertebrae

According to movement

(1) **Movable or True:** The vertebrae which can freely move and united with the adjacent movable vertebrae by means of a fibro-elastic substance

(inter-vertebral disc) is known as movable vertebrae. All the vertebrae except sacral vertebrae are movable vertebrae.

(2) **Immovable or Fixed or False:** These vertebrae which are rigidly fused to form single bone e.g. Sacral vertebrae.

According to their mode of formation the vertebrae are

(a) **Typical:** Typical vertebra presents a body, arch and certain processes. Examples - Third cervical vertebra.

(b) **Atypical: (A = not):** It is a vertebra which deviates from the general plan of construction of the typical vertebrae. In this kind of vertebrae there is a total suppression of some of the typical parts of vertebrae (in Atlas the body and the dorsal spinous process are absent) or addition of a part (in axis additional odontoid process is present).

a. Comparative Study of Cervical Vertebrae

The vertebral segment of the neck region is called as cervical vertebrae. They are seven in number (in all mammals) placed serially one behind the other. They are massive, quadrangular, and longer than other region of the vertebrae. Their dorsal spinous process is inclined, articular processes are well developed and transverse processes are divided into upper and lower part. In between the two parts of transverse process a foramen, foramen transversarium is present which gives passage to the vertebral artery. The 1st (atlas) and 2nd (axis) vertebrae are highly modified and comes under atypical vertebrae. 3rd, 4th 5th, 6th and 7th cervical vertebrae are typical type.

First Cervical Vertebra or the Atlas of Ox

It is an atypical vertebra as the body and vertebrae the dorsal spinous process are absent. The dorsal spinous process is in the form of a strong ring, from which two curved plates, the wings project laterally. The atlas comprised of two lateral masses, a bony ring and wing (Fig. 2.22).

The bony ring encloses a very large vertebral foramen, and consists of two lateral masses connected by dorsal and ventral arches. Lateral masses present two deep oval cranial articular cavities which articulates with the condyles of the occipital bone of skull. These cavities are separated by a wide notch above and a narrow notch below. Each cavity is partially divided into dorsal and ventral parts by a non-articular area. The caudal articular surfaces are confluent on the ventral arch, but are widely separated dorsally. These are flattened articular structures continued into the vertebral canal, forming an extensive area for the dens of axis, called *Fovea dentis*. The dorsal arch is convex dorsally and concave ventrally. *Muscle rectus capitis dorsalis minor* originates from the dorsal arch of the atlas. A large median dorsal tubercle is present dorsally on dorsal arch. Two intervertebral/ lateral foramen one on each side perforates the dorsal arch near its cranial margin. These foramens are connected with alar foramen (which perforates the wing) by a short alar groove. The cranial border of the dorsal arch has a wide deep notch, while, the caudal has only a narrow and shallow notch. The ventral arch is thicker, narrower and less curved than the dorsal arch ventral tubercle is present at the middle of its lower surface for attachment of tendon of *Longus colli muscle*. A transversely flat articular surface, the *Fovea dentis* is present on the dorsal surface of atlas for the *dens* of the axis. A transverse rough area for the *Ligamentum dentis* (odontoid ligament) is located in front of *Fovea dentis*. *Muscle rectus capitis ventralis minor* originates from the ventral surface of the lateral mass of the atlas close to the caudal articular surface. *Muscle obliquus capitis posterior* inserts on the dorsal surface of the wing of the atlas. Whereas, The *Muscle oblique capitis anterior* originates from the cranial border and ventral surface of the wing of the atlas.

The wings are slightly curved plates which project laterally and slightly downward and backward from the lateral masses. These are modified form of transverse processes of the other vertebrae. The border of the wing is slightly thicker, rough and straight and gives attachment to the *longissimus capitis et atlantis* muscle. A cavity the *fossa atlantis/ atlantal fossa* is present between the central part of the wing ventrally and the lateral mass. Each wing is perforated by two foramina. The Alar foramen perforates the wing cranially and provides passage to the first cervical spinal nerve. It communicates laterally with atlantal fossa. The lateral vertebral foramen leads medially into vertebral canal.

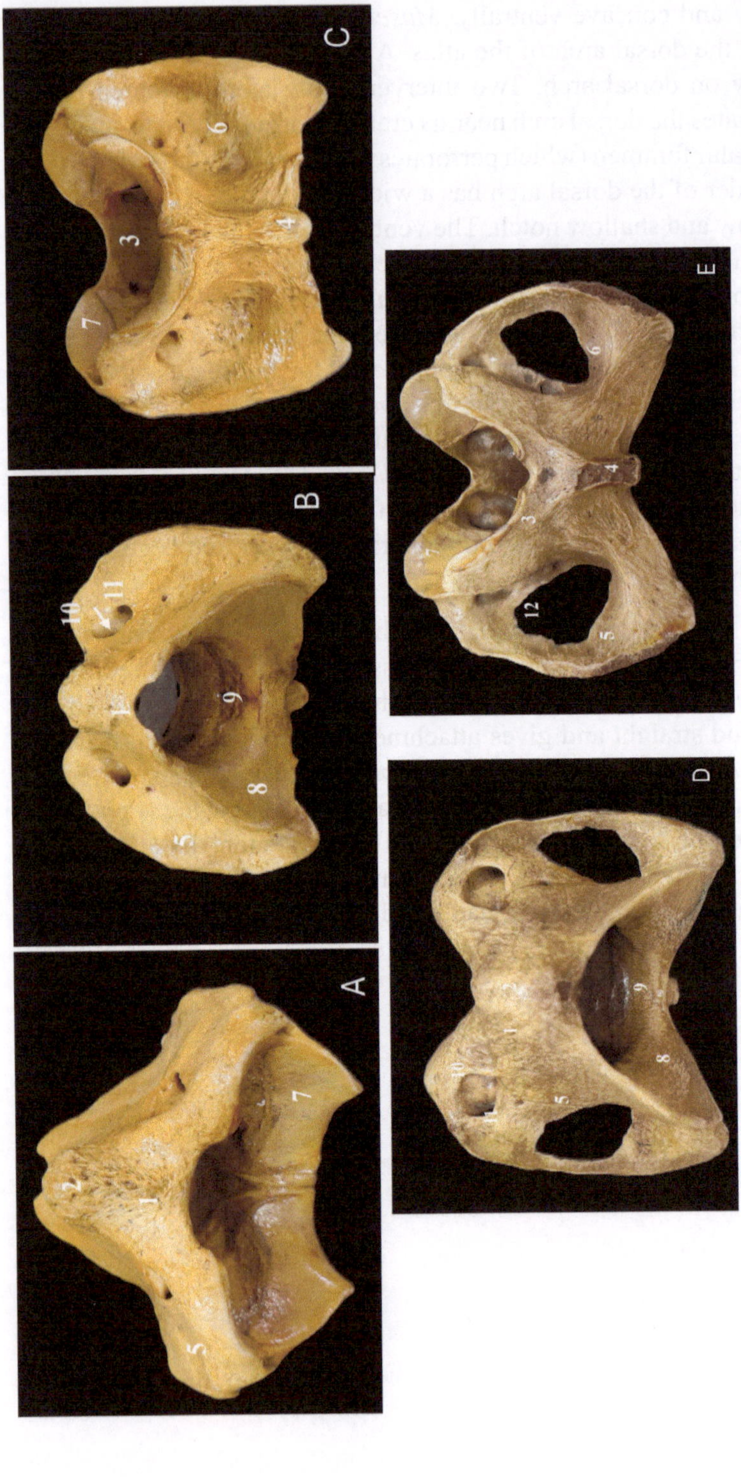

Fig. 2.22: Photograph showing the first cervical vertebra (atlas) of Ox (A-C) and Horse (D and E): A. and D. Dorsal view, B. Cranial View, C. and E. Ventral view, 1. Dorsal arch, 2. Dorsal tubercle, 3. Ventral arch, 4. Ventral tubercle, 5. Wing, 6. Fossa atlantis, 7. Cranial articular facet, 8. Caudal articular facet, 9. Fovea dentis, 10. Foramen alar, 11. Lateral vertebral foramen, 12. Transvers foramen

Table 2.17: Comparative table on Atlas of Ox, Horse, Dog and Pig

Characters	Ox	Horse	Dog	Pig
Ventral arch	Narrow	More prominent ventral tubercle	Narrow from before backward, and bears a small tubercle caudally	Ventral tubercle is long, compressed laterally and project back under the axis
Dorsal arch	Broad and convex	More convex than ox, Dorsal tubercle is indistinct	Strongly convex and rough centrally, dorsal tubercle is absent	Dorsal tubercle is larger than ox
Cranial articular caviies	Deeply excavated, divided into small dorsal and large ventral part by a non-articular area	Same as in ox	Same as in ox	Same as in ox
Caudal articular surface	Flattened	Somewhat saddle shaped	Moderately concave	Partly divided into two parts by a lateral projection on each side
Wings	Less curved than horse.	More curved than in the ox. They are thinner than ox but their borders are thick.	Wide, flattened, and almost horizontal and proceed further backward than ox and horse	Narrow, Flattened, horizontally directed and bears a caudal tuberosity
Foramen transversarium	Absent	Additional foramen presents in the caudal part of the dorsal surface of the wings	Same as in horse but canal like	Located on the caudal border of the wing
Lateral vertebral foramen	Located cranio- medially	Same as in ox	Same as in ox	Same as in ox
Alar foramen/ notch	Present in the wing	Same as in ox	Alar notches on the cranial border instead of the alar foramen	Same as in ox
Fovea dentis	Extensive	Transversely concave	Shallow	Same as in dog
Fossa atlantis	Moderately concave	Deeper than ox	Shallower than ox	Deeper than dog

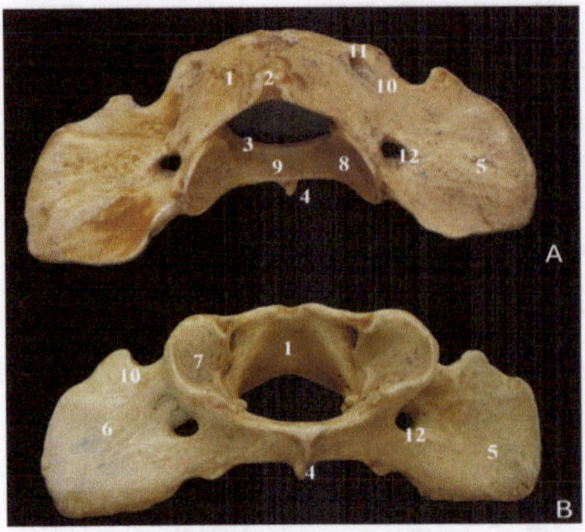

1. Dorsal arch
2. Dorsal tubercle
3. Ventral arch
4. Ventral tubercle
5. Wing
6. Fossa atlantis
7. Cranial articular facet
8. Caudal articular facet
9. Fovea dentis
10 Alar notch
11 Lateral vertebral foramen
12. Transvers foramen

Fig. 2.23: Photograph showing the first cervical vertebra (atlas) of Dog A. Caudo-dorsal view, B. Cranio-ventralview

Second Cervical Vertebra or the Axis of Vertebra Ox

Axis is second atypical cervical vertebra and also known as *Vertebra Prominance/ Vertebra Denoteta / Epistropheus* because of presence of the dens or odontoid process, which project form cranial part of the body. The cylindrical body of the axis is longest among all the vertebrae. Its cranial extremity is characterized by presence of dens. The dens/ odontoid process is wide, articular prominence of the axis. A convex articular surface on its ventral surface articulates with the ventral arch of the atlas. The dorsal surface of the dens is deeply concave and has got a depression for the attachment of the *ligamentus dentis*. The caudal extremity of the body is similar to the other vertebrae and has a usual cotyloid cavity (Fig. 2.23).

Cranial articular processes are modified and flanks from dens on either side of it. These processes have saddle shaped articular surfaces and are confluent ventrally with that of dens. The caudal articular processes are typical. In the young animals the arch has a notch on either side of its cranial border which later on ossified and is converted into a lateral vertebral foramen by a ligament (Fig. 2.23). The transverse processes are small, single, and project backward. The foramen transversarium is small or sometimes absent. The dorsal spinous process is prominent and projects a little infront. Its height and thickness gradually increase from before backward. Its dorsal/ free border is rough. The caudal border of dorsal spinous process abruptly descends downward. The ventral spine resembles that of the typical vertebrae. The *muscle rectus capitis dorsalis major* originates from the cranial border of the spine of the axis and extends from it to the occipital bone. *Muscle obliquus capitis posterior* origins from the lateral surface of the spine and the posterior articular process of the axis (Fig. 2.23).

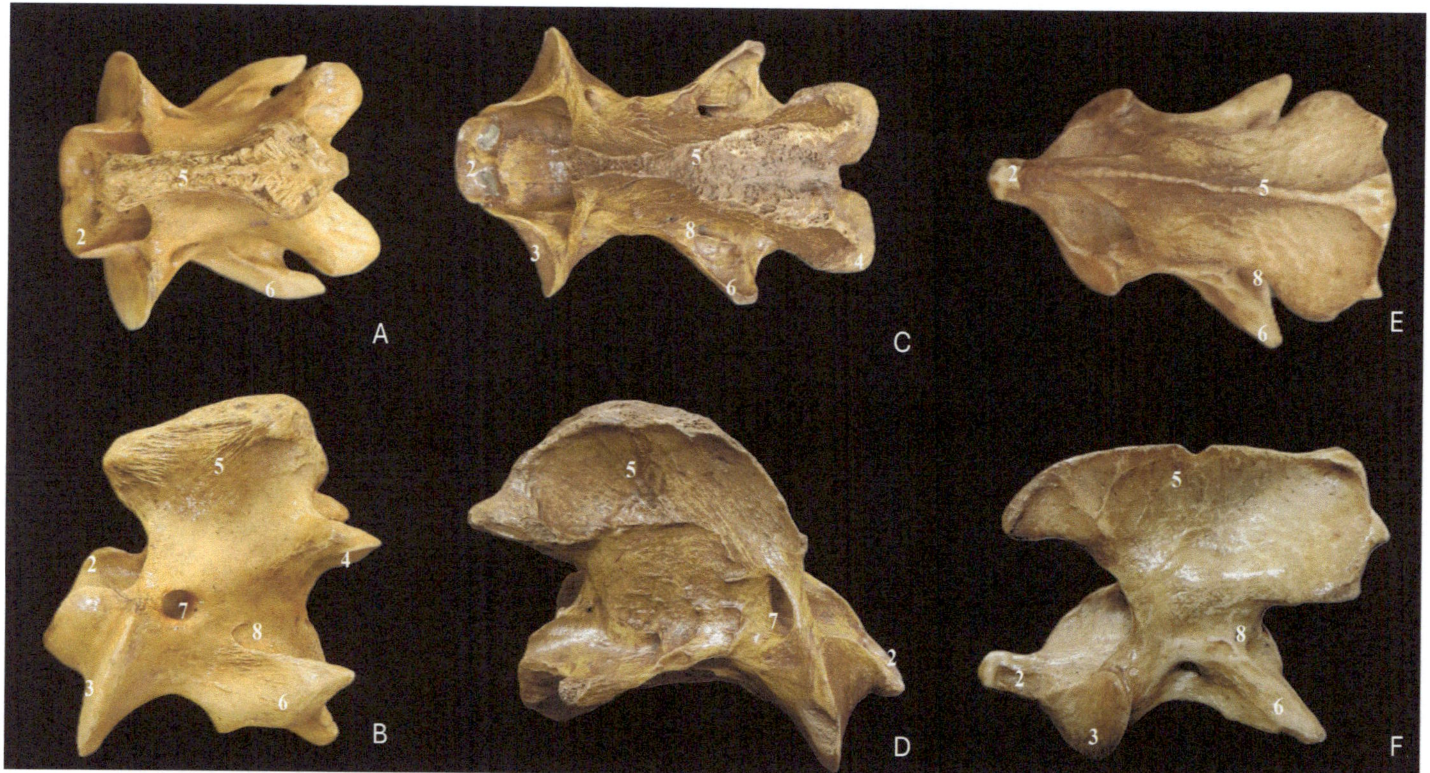

Fig. 2.24: Photograph showing the second cervical vertebra (Axis) of Ox (A and B), Horse (C and D) and Dog (E and F) 1. Body, 2. Dens, 3. Cranial articular process, 4. Caudal articular process, 5. Dorsal spinous process, 6. Transverse process, 7. Lateral vertebral foramen , 8. Trans vers foramen

Table 2.18: Comparative table on Axis of Ox, Horse, Dog and Pig

Characters	Ox	Horse	Dog	Pig
Shape/ Size	Body is cylindrical, shorter than horse and have distinct ventral spine	Body is about two times longer than ox.	Flattened dorsa-ventrally, specially infront	Same as in dog
Arch	Similar to typical cervical vertebra	Similar to typical vertebra	Overhang the cranial and caudal end of the body	Overhang at caudal end of the body
Ventral surface	Wide, and is divided by a median crest into the two fossae	Same as in ox	Same as in ox	Same as in ox
Odontoid process/ Dens	Large, stout, tooth-like process with a deeply concave dorsal surface. ventral articular surface od dens is confluent with cranial articular surface.	Longer, conical, narrower, and more pointed and scoop/ spatula shaped on its upper aspect. Ventral articular surface is similar to ox.	Rounded and relatively long, reaching almost to the occipital bone and is inclined a little upward, articular surface is oblique and convex. Ventral articular surface is separated from cranial articular surface	Thick cylindrical rod shaped. Similar to that of dog.
Dorsal spinous process	Rectangular bony plate, which does not bifurcate caudally, remain separated and does not confluent with caudal auricular process	Longer and stronger than in the ox. Its free border is rough and thicker caudally and continued with the caudal articular processes by two ridges.	Thin and of moderate height, but very long. It is prolonged forward so as to overhang the dorsal arch of the atlas and is terminated behind by a tuberosity which is connected by two crests with the caudal articular processes.	Large, directed dorsal and caudal but does nor confluent with caudal auricular process
Fovea dentis	Extensive	Transversely concave	Shallower than ox	Shallower than ox
Transverse process	Project caudally beyond the level of caudal extremity	Short, do not project beyond the level of caudal extremity	Single, pointed, directed backward and outward	Very small, directed ventrally

Foramen transversarium	Present but smaller than horse and pig	Longer than that of the ox	Present smaller than pig	Larger than other species
Intervertebral foramen	Circular	Oval and is very close to the cranial border of the arch	Cranial notches are large and are never converted into lateral vertebrae foramina	Bridged cranially and laterally by bony plate

Typical Cervical Vertebrae (3^{Rd} to 7^{Th}) of Ox

Third to seventh cervical vertebra having characters of typical vertebrae. Each vertebra presents body, arch and processes. The size of the body progressively reduced from cranial to caudal (Fig. 2.24). In 7th cervical vertebra the body is short and has a small facet on either side of the cotyloid cavity for the head of first rib (Fig. 2.29). The body presents dorsal and ventral surfaces, cranial and caudal extremities and dorsal and ventral spinous processes. The dorsal surface of the body is slightly rough in the centre and gives attachment to the dorsal longitudinal ligament. The ventral surface presents a short median ventral spinous process which separates two concave areas. This ventral spine becomes indistinct or absent in 6th and 7th cervical vertebrae. The cranial extremity of the body has an oval articular head. It is strongly convex, wider above than below. The caudal extremity is larger and has a large circular cotyloid cavity. The *Muscle longus colli* lies on the ventral surface of the all cervical and first six thoracic vertebrae. The cervical part of this muscle inserts to the bodies of the cervical vertebrae and ventral tubercle of the atlas. The dorsal spinous processes are well developed, and increase in height from before backward (3rd to 7th). They are directed upward and forward. The summit of the dorsal spinous process of third vertebra is usually bifid. The spinous process of 6th cervical is longer about two inches in length and more strongly inclined forward (Fig. 2.26). In 7th cervical vertebra the dorsal spinous process is nearly vertical and is about four or five inches in height (Fig. 2.27). This bone (7th cervical) receives the additional name of vertebra prominences because of the greater height of the spine.

The arch is perforated by a foramen on either side which communicate with foramen transversarium. The vertebral notches of the arch are large (Fig. 2.24). The articular processes are four in number, two cranial and two caudal. A bony plate connects the articular process of the same side. These processes are large with extensive, oval and slightly concave articular surfaces. The cranial ones are dorso-medially directed and the caudal processes are ventro-laterally directed. The transverse processes are large and plate like. The *muscle multifidus cervicis* lies on the dorsal surface of the arches of the last five cervical vertebrae. The *muscle longissimus capitis et atlantis* consists of two narrow bands extending along the cervical vertebrae just above the articular processes except the first two or three. They occupy the space between the articular and transverse processes on the lateral surface of the cervical vertebrae. Each transverse process arises from two roots, one from the arch and another one from the body. Foramen transversarium is formed between these two roots (Fig. 2.26). Largest foramen transversarium is present in 6th cervical vertebra, whereas, in 7th cervical it is absent (Fig. 2.29). Vertebral vessels and a nerve pass from this foramen. Canalis transversarius is formed by aggregation of these foramina. Each transverse process divides laterally into cranial and caudal branches which are thickened and rough for muscular attachment (Fig. 2.24). In 6th cervical vertebra the transverse process has two parts, the upper is short and tuberous, while the lower one is a large, thick, quadrilateral and almost sagittal plate, directed ventrally (Fig. 2.26). The transverse process is undivided and short and thick in 7th cervical vertebra (Fig. 2.29). The *Ventral scalenous muscle* originates from the transverse processes of the fourth to the sixth cervical vertebrae. Similarly, the *dorsal scalenous muscle* having its origin on the transverse process of the sixth and possibly the fifth and fourth cervical vertebrae. The *Muscle rectus capitis ventralis major* originates from the ventral branch of the transverse processes from second to the sixth cervical vertebrae. The *splenius muscle* take insertion to the transverse processes of the first three cervical vertebrae.

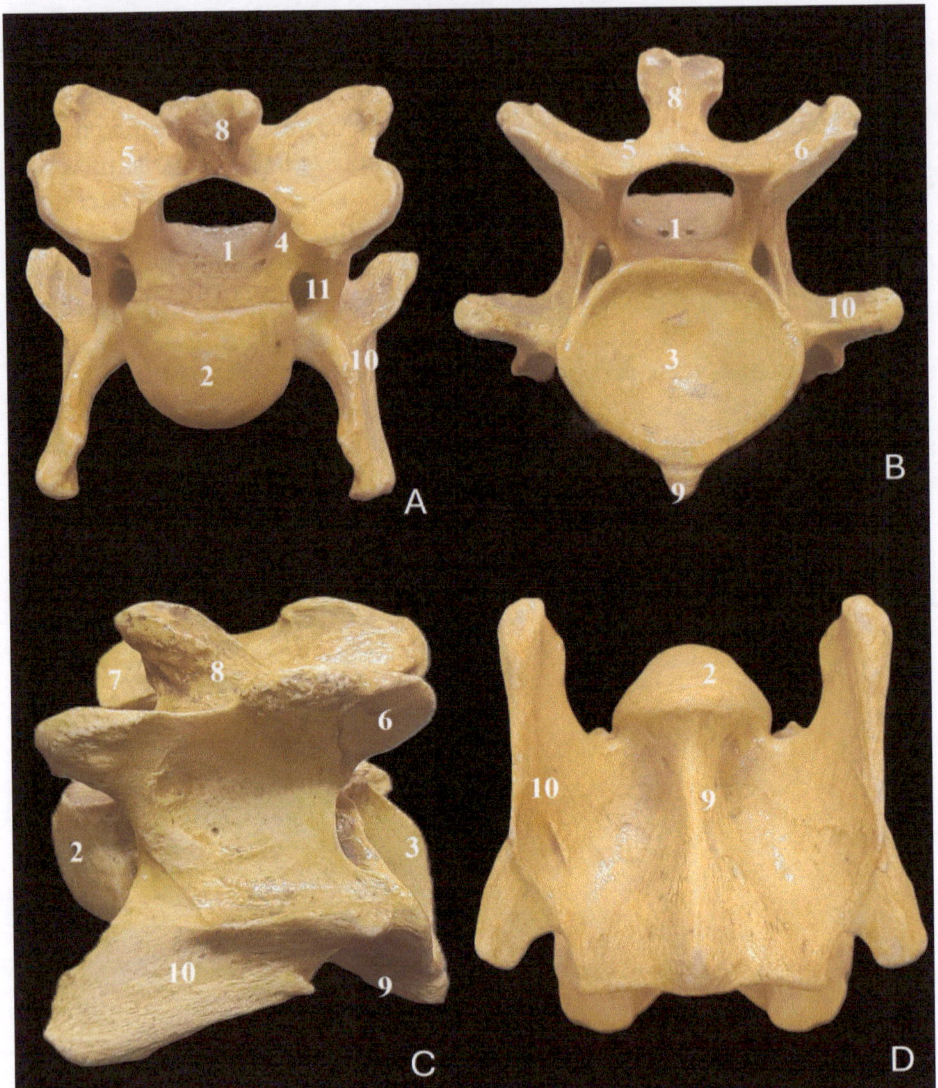

Fig. 2.25: Photograph showing the typical cervical vertebrae of Ox A. Cranial view B. Caudalview C. Lateral view D. Ventral view, 1. Body, 2. Head, 3. Cotyloid cavity, 4. Pedicle of arch, 5. Lamina of arch, 6. Cranial articular process, 7. Caudal articular process, 8. Dorsal spinous process, 9. Ventral spinus process, 10. Transvers process, 11. Foramen transversarium

Table 2.19: Comparative table on 3rd, 4th and 5th Cervical Vertebrae of Ox, Horse, Dog and Pig

Characters	Ox	Horse	Dog	Pig
Body	Long as compared with those of other vertebrae but Smaller than horse	Much longer and larger than in the ox.	Relatively longer than in ox and pig. The length of body diminishes from first to last (i.e. 3rd to 7th) and are compressed dorsoventrally	Short and wide body is elliptical in cross section
Cranial extremity of the body	Convex, wider above than below	More convex than ox	Oblique, moderately convex than ox	Slightly convex or flattened
Caudal extremity of body	Concave cotyloid cavity	More concave than ox	Oblique and slightly concave	Slightly concave
Cranial articular process	Large, oval and dorso-medially directed	Oval in shape but larger than in the ox	Same as in ox	Slightly convex from side to side and concave dorso-ventrally
Caudal articular process	Oval and ventro-laterally directed	Larger than in the ox	Same as in ox	Slightly concave
Dorsal spinous process	Well developed, directed upward and forward, and increase in height from before backward.	Form of a low crest, which widens behind, and is connected by ridges with the caudal articular processes.	Third has the form of a long low crest, in the remainder it is higher, blunt pointed, and inclined forward.	3-5 increase in height cranial one inclined caudally and caudal one cranially inclined
Ventral spinous process	Distinct	Same as in ox	Distinct, become more pronounced caudally	Not present
Transverse process	Divided into two parts horizontal, dorso-ventrally flattened dorsal tubercle, and cranio-ventrally directed, elongated ventral tubercle	Same as in ox	Project ventrally and backward, and divide into two branches; of these, the cranial one is thin, and the caudal is thick and tuberculate at its free end.	Increase in size from 3 to 5, Divided into two branches (dorsal branch is short and caudo lateral, ventral is quadrilateral plate directed ventrally)
Foramen transversarium	Present, size gradually increased from 3-5	Same as in ox	Same as in ox	Same as in ox

Note: 1. Horse: A body plate connects the cranial and caudal articular processes of the same side in the fifth cervical vertebra. In the case of the fourth cervical vertebra this bony plate extends only three-quarter distance forward. In the third cervical vertebra this plate extends only half the distance.

2. Pig: The pedicle is perforated by foramen in addition to the intervertebral foramen.

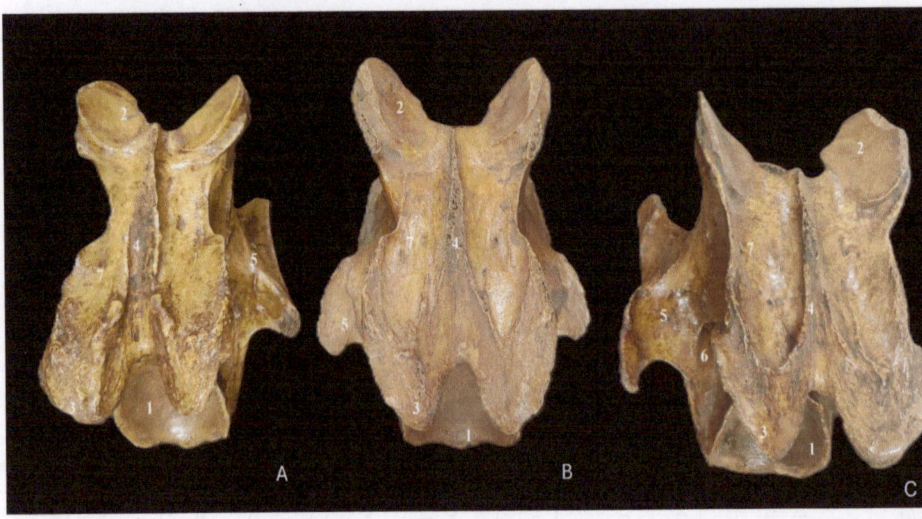

Fig. 2.26: Photograph showing the cervical vertebrae of Horse: A. 3rd cervical, B. Fourth cervical and C. Fifth cervical 1. Cotyloid cavity, 2. Cranial articular process, 3. Caudal articular process, 4. Dorsal spinous process, 5. Transverse process, 6. Foramen tranversarium, 7. Bony plate connecting the cranial and caudal articular processes of same side

Comparative Study of 6th Cervical Vertebra of Ox

1. The dorsal spinous process is longer and more strongly inclined forward and may be about two inches in length.
2. The transverse process has two parts the upper part is short and tuberous, while the lower one is a large, thick, quadrilateral and almost sagittal plate, directed ventrally.
3. The foramen transversarium is largest.
4. The ventral spine is absent.

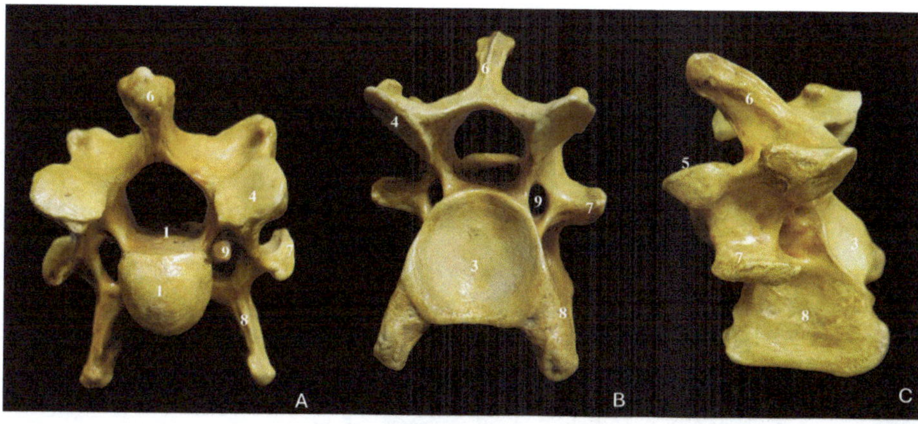

Fig. 2.27: Photograph showing the 6th cervical vertebra of Ox A. Cranial view, B. Caudal view and C. Lateral view: 1. Body, 2. Head, 3. Cotyloid cavity, 4. Cranial articular process, 5. Caudal articular process, 6. Dorsal spinous process, 7. Dorsal tubercle of transverse process, 8. Ventral plate of transverse process, 9. Foramen tranversarium

Table 2.20: Comparative table on 6th Cervical Vertebrae of Ox, Horse, Dog and Pig

Characters	Ox	Horse	Dog	Pig
Body	Short but wide	Same as in ox	Relatively longer than in ox and pig	Short and wide
Dorsal spinous process	Longer and more strongly inclined forward and may be about two inches in length	Half inch or more in height.	Blunt pointed, and inclined forward.	3rd is directed caudally, 4th is directed vertical, and 5th is directed cranially
Caudal articular process	Large	Caudal articular processes are shorter, thicker and further apart	Same as typical vertebra	Same as typical vertebra
Ventral spine	Absent	Small and is less prominent caudally	Absent	Absent
Transverse process	Two Parts, upper is short and tuberous, while the lower one is a large, thick, quadrilateral and almost sagittal plate, directed ventrally	Three branches Dorsal, Cranial; and Caudal. the caudal part is a thick, almost sagittal plate, which forms with its fellow and the body a wide ventral groove; the dorsal and cranial branches are shorter and thicker than the typical vertebrae.	Two parts; one is ventro laterally directed extensive quadrilateral plate shaped with a ridge on its medial surface; the other part is short and blunt, and is directed outward and a little backward and upward.	Same as typical vertebrae except its size is more than 5th

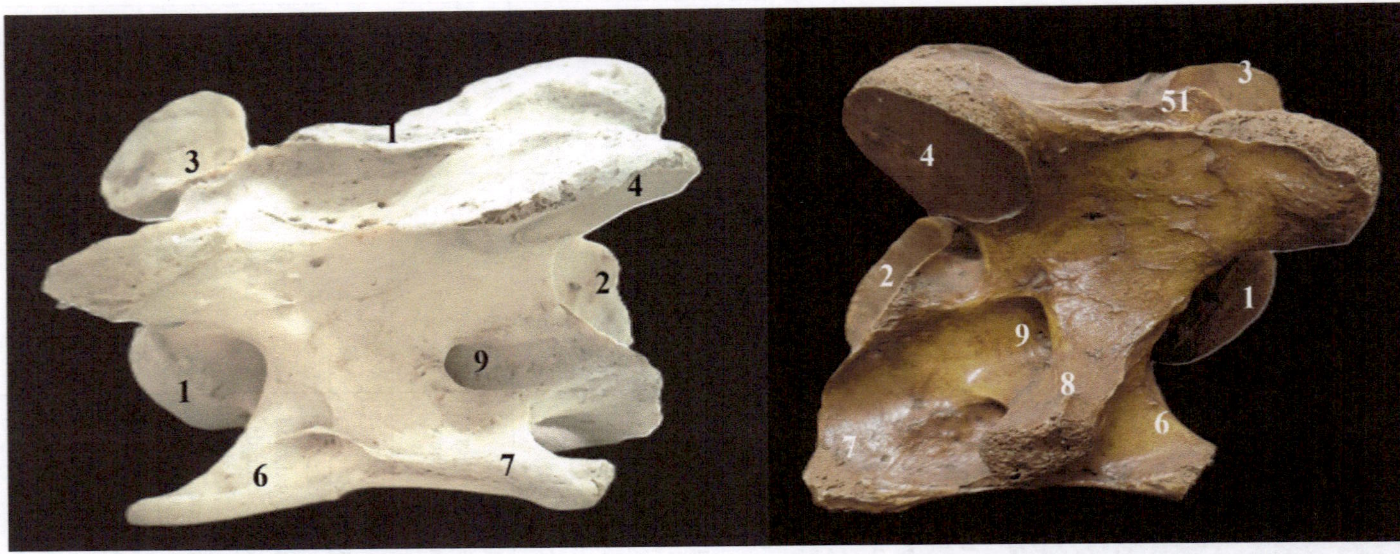

Fig. 2.28: Photograph showing the cervical vertebra of Horse **A.** Third cervical, **B.** Sixth cervical 1. Head, 2. Cotyloid cavity, 3. Cranial articular process, 4. Caudal articular process, 5. Dorsal spinous process, 6. Cranial part of transverse process,7. Caudal part of transverse process, 8. Lateral part of transverse process, 9. Foramen tranversarium

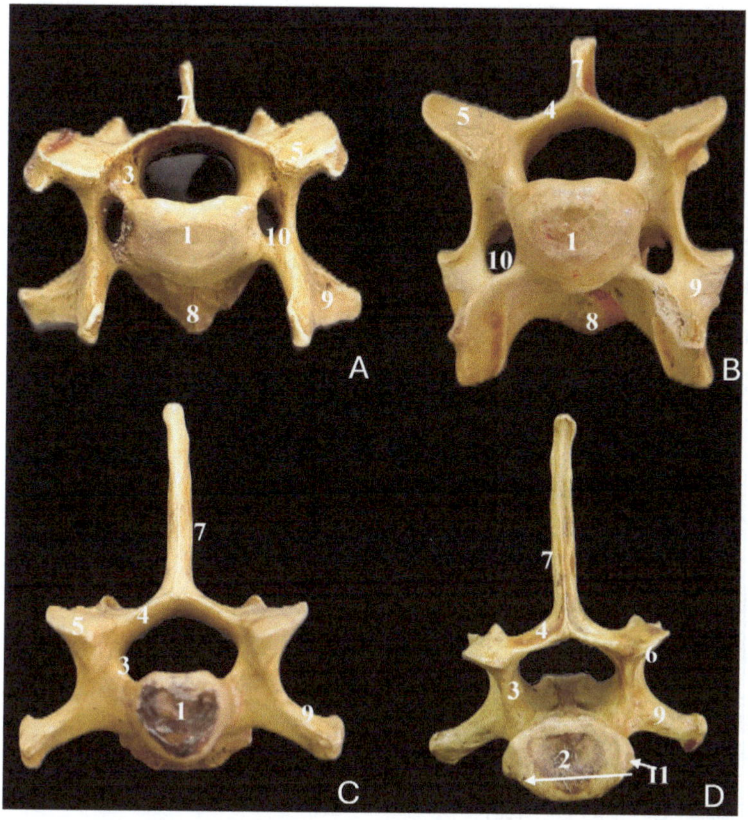

Fig. 2.29: Photograph showing the cervical vertebrae of dog **A.** 5th cervical (cranial view), **B.** 6th cervical (cranial view), **C** and **D** 7th cervical vertebrae (cranial and caudal view), 1. Head, 2. Cotyloid cavity, 3. Pedicle of arch, 4. Lamina of arch, 5. Cranial articular process, 6. Caudal articular process, 7. Dorsal spinous process, 8. Ventral spinous process, 9. Transverse process, 10. Foramen transversarium, 11. Costal facets

Comparative Study of 7th Cervical Vertebra of Ox

1. The dorsal spinous process is nearly vertical and is about four or five inches in height.
2. The body is short and has a small facet on either side of the cotyloid cavity for articulation with the facet present on the head of first rib.
3. The arch and notches are large.
4. The transverse process is undivided and is short and thick.
5. Foramen transversarium is absent.

Table 2.21: Comparative table on 7th Cervical Vertebrae or Vertebra Prominence of Ox, Horse, Dog and Pig

Characters	Ox	Horse	Dog	Pig
Size of body	Short, caudally bears a pair of facets for head of the first rib	Comparatively longer than that of ox caudally bears a pair of facets for head of the first rib	Short caudally bears a pair of facets for head of the first rib	Short and Flat caudally bears a pair of facets for head of the first rib
Cranial articular process	Small	Wider and longer than caudal ones	Similar to 3-5th	Similar to 3-5th
Caudal articular process	Small	Longer than ox, large plate like articular surface	Similar to 3-5th	Similar to 3-5th
Dorsal spinous process	Nearly vertical and is about four or five inches in height	Pointed, an inch or more in height	Similar to 3-5th	About 10 cm in height
Transverse process	Undivided, short and thick	Undivided	Undivided	Undivided
Foramen transversarium	Absent	Absent	Absent	Present on arch along with another foramen
Ventral spine	Absent	Replaced by a pair of tubercles.	Present	Absent

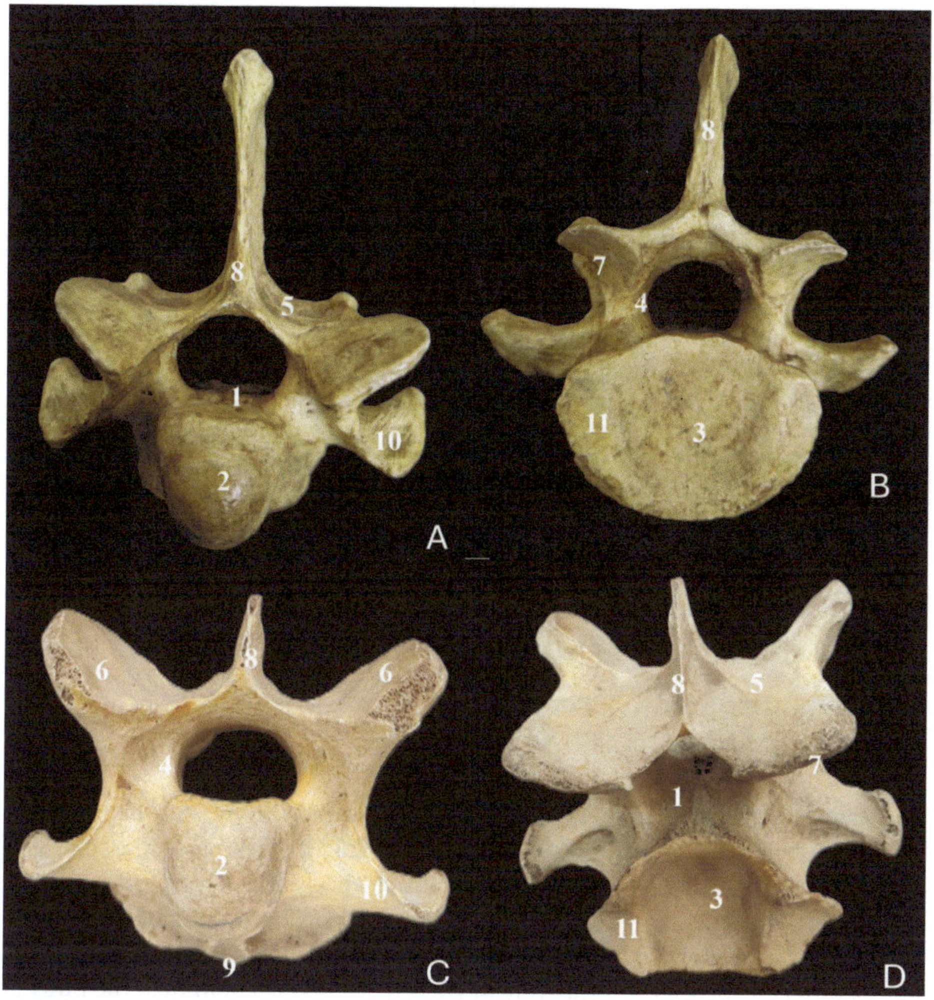

Fig. 2.30: Photograph showing the 7th cervical in Ox (A and B) and Horse (C and D). A and C: Cranial view, B and D caudal view

1. Body
2. Head
3. Cotyloid cavity
4. Pedicle of arch
5. Lamina of arch
6. Cranial articular process
7. Caudal articular process
8. Dorsal spinous process
9. Ventral tubercle
10. Transverse process
11. Costal facet

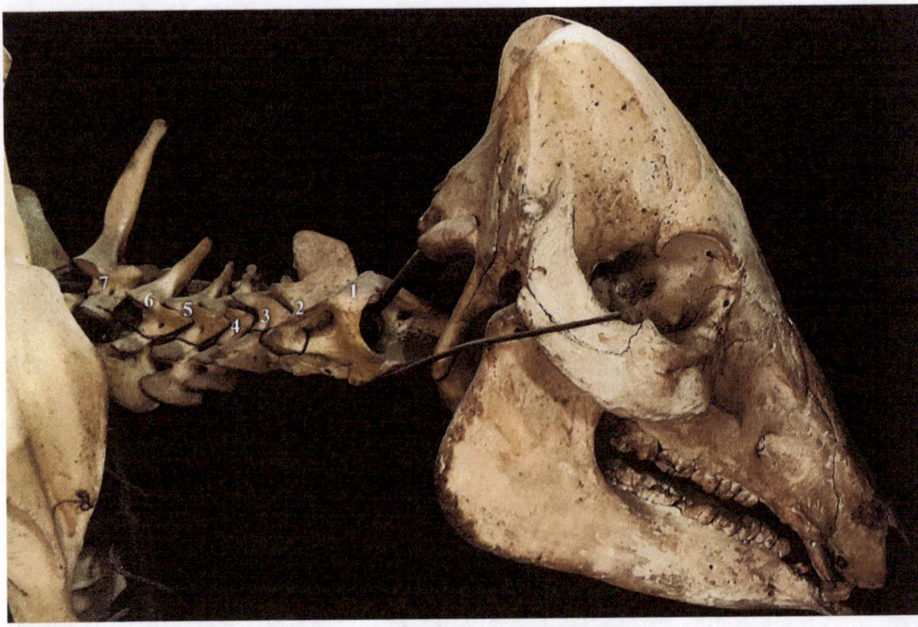

Fig. 2.31: Photograph showing the cervical vertebrae of Pig (1ST TO 7TH)

b. Comparative Study of Thoracic Vertebrae

Thoracic Vertebrae of Ox

These vertebrae form the bony roof of the thoracic cavity along with the proximal end of the ribs. They are 13 in number (Fig. 2.35). These are characterized by long dorsal spinous process and cylindrical body. Each vertebra presents three pairs of facets, two pairs on the either end of the body situated on the dorsolateral aspect of the body for the head of the rib and one pair on the ventral side of transvers process for tubercle of rib (Fig. 2.32). The length of the body is gradually increased. The ventral spinous process is prominent. The cranial and caudal extremity of the body are flattened corresponding to the inter-vertebral disc. The dorsal spinous processes are very long and directed backward. The dorsal spinous process of first thoracic vertebra is about nine inches in height. In the next two or three vertebrae these processes had a highest dorsal spinous process. Behind them there is gradual decrease in their height (Fig. 2.31 and 2.35). Dorsal spinous process of 3rd and 4th vertebrae is maximum in length. The backward slope of the dorsal spinous process is slight at first. It increases to the tenth vertebra whereas the spinous process of the thirteenth vertebra is vertical and lumbar in character (Fig. 2.35). The summit of the dorsal spinous process is pointed in the first and thickens on those which are further back. The summits of the last seven dorsal spinous processes are bifid.

Either side of the arch is pierced by additional lateral vertebral foramen. The arch has cranial shallow and caudal deep vertebral notches. Large intervertebral foramen is formed between corresponding notches which is often divided by a bony ridge. The transverse processes are short, thick and undivided. Each of them bears a ventral articular facet for articulation with the tubercle of the corresponding rib. The transverse process diminishes in size from before backward (1st to 13th). The costal facets (present on body and transverse process) become gradually smaller from the first to the last vertebra. The upper non-articular part of the transverse process is thick and strong. It carries the mamillary process from eighth vertebra to eleventh one. The mamillary process is again present on the transverse process of the thirteenth thoracic vertebra. The mamillary process fused with cranial articular process to form combined *mamilloarticular process*.

The articular processes are in the form of oval facets situated at the corresponding aspect of the base of the arch. The cranial facet is directed cranio-dorsally and caudal facets are directed caudo-ventrally. The cranial facets are tangentially oriented to the vertebral arch whereas, the caudal facets are sagittally oriented (Fig. 2.31). The last thoracic vertebra bears only two facets at cranial end of the body for last pair of rib.

In first thoracic vertebra, the body is wide and flattened dorso-ventrally. The cotyloid cavity is deeper than any other thoracic vertebra. Two large costal facets are present on either side of extremities of body. The arch is large and strong, and has large notches. The articular processes are much larger than those of other thoracic vertebrae (Fig. 2.33). The length of dorsal spinous process is nearly nine inches. The transverse processes of the first two thoracic vertebrae give origin to the *muscle longissimus capitis et atlantis*. Similarly, the transverse processes of the first ten thoracic and the articular processes of the last five cervical vertebrae give origin to the *muscle complexus (Muscle semispinalis capitis)*.

Note: the thoracic vertebra whose dorsal spinous process is nearly perpendicular to the long axis of that bone is known as Anticlinal or diaphragmatic vertebra.

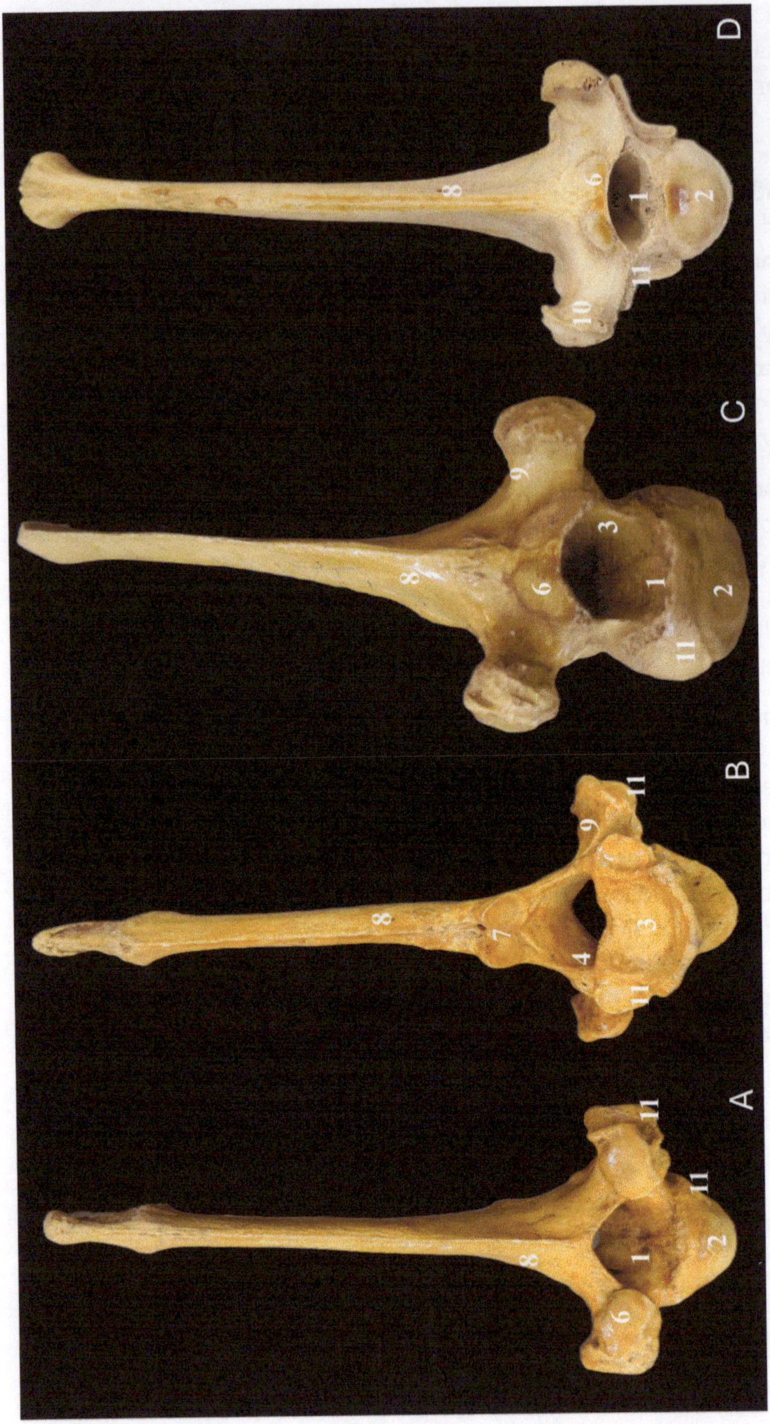

Fig. 2.32: Photograph showing the thoracic vertebrae of Ox (A. 1st thoracic cranial view, B. 1st thoracic caudal view), C. Thoracic vertebrae, D. Thoracic vertebra of horse, 1. Body, 2. Cranial extremity of the body, 3. Caudal extremity of the body, 4. Pedicle of arch 5. Lamina of arch, 6. Cranial articular process, 7. Caudal articular process, 8. Dorsal spinous process, 9. Transverse process, 10. Mamillary process, 11. Costal facet

Table 2.22: Comparative table on Thoracic vertebrae of Ox, Horse, Dog and Pig

Characters	Ox	Horse	Dog	Pig
Number	13	18	13	14/15
Dorsal Spinous process	3rd and 4th have maximum length. Backward slope increases up to the 10th vertebra whereas the spinous process of the thirteenth vertebra is vertical and lumbar in character. The Summits of the last seven dorsal spinous processes are bifid.	Large, narrow, and slopes upward and backward. The summit is rounded. Increase in height upto first four then gradually diminishes upto 15th or 16th . in 15th and 16 vertebra this process is vertical and in 17th and 18th vertebra equal in height which inclines forward	First 3-4 equal in length then gradually decreased upto 10 and then equal.11-vertical, last 2 inclined cranially	From 2nd to 9th - height decreases gradually and inclined caudally. From 10th to 14th or 15th - equal height, 10th is diaphragmatic, 11th vertical, 12-14 inclined cranially
Dorsal Spinous process of 1st thoracic	The backward slope of the dorsal spinous process is slight at first. The summit of the dorsal spinous process is pointed in the first and thickens on those which are further back.	The dorsal spinous process of the first thoracic vertebra is much shorter than that of ox and tapers to a point. Its height is usually four to five inches	First 3-4 equal in length	Broad, very high, inclined cranially
Anticlinal vertebra	13th	16th	10th	12th
Costal facets	Three pairs, two pairs are present on the body (cranially and caudally) and one pair is present on the transverse process	Same as in ox	Three pairs` Caudal pair of facets are absent in last 2 or 3 vertebrae	Same as in ox
Transverse process	Short, thick and undivided, size reduced from 1-13	Same as in ox	Short and facets become smaller from 1-13	Same as in ox
Tubercular facet	Present on ventral aspect of transverse process, size reduced from 1-13	Same as in ox	Large and concave at cranial end and small and slightly concave caudally	In last five or six vertebrae facet for head and tubercle get fused

Characters	Ox	Horse	Dog	Pig
Mammillary process	Present between transverse and cranial articular process of eighth to eleventh and 13th vertebra	Distinct in last 4 or 5 vertebrae	Present except the cranial end of series	Present except the first 2 thoracic
Accessory process	Not present	Not present	Present in last three vertebrae	Occurs in caudal part of the region (last 4)
Intervertebral foramen	Often divided by bony ridge	Not divided	Not divided	Divided into dorsal and ventral parts

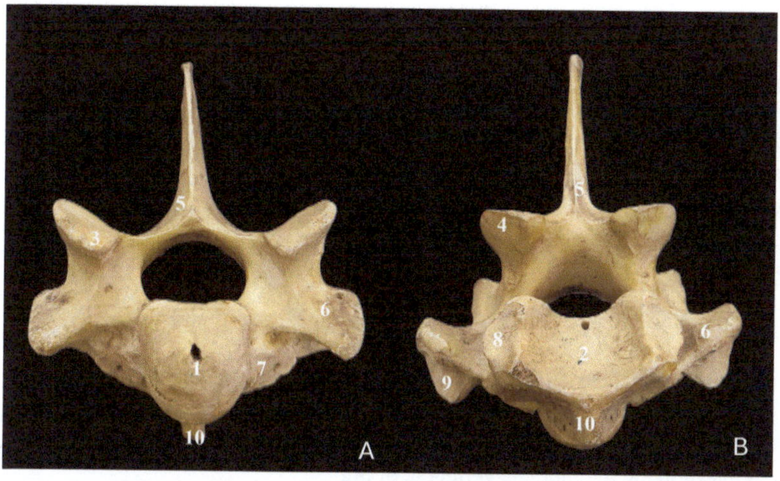

Fig. 2.33: Photograph showing the 1st thoracic vertebra of Horse **A.** Cranial view, **B.** Caudal view: 1. Cranial extremity of the body 2. Caudal extremity of the body, 3. cranial articular process, 4. caudal articular process, 5. dorsal spinous process, 6. transverse process 7, Costal facet on head, 8. Costal facet on cotyloid cavity, 9. Costal facet on tubercle, 10. Ventral spinous process

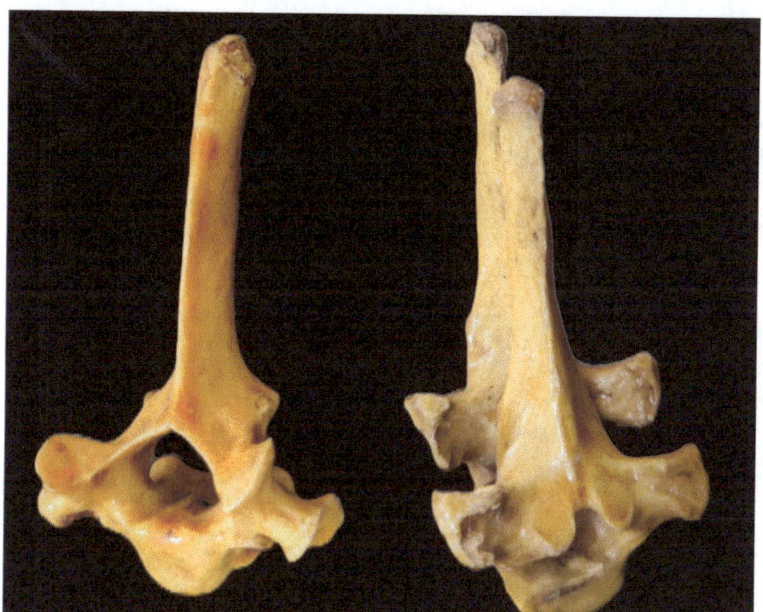

Fig. 2.34: Photograph showing the thoracic vertebra of dog **A.** cranialviewof 1stthoracic, **B.** cranial view of other thoracic: 1. Head, 2. Cranial articular process, 3. Dorsal spinous process, 4. Transverse process 5. Costal facet on head, 6. Costal facet on tubercle

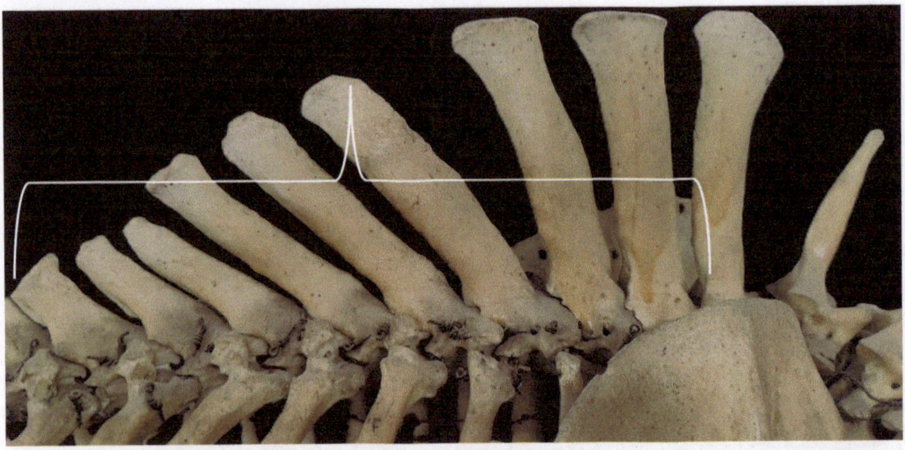

Fig. 2.35: Photograph showing the thoracic vertebrae of pig

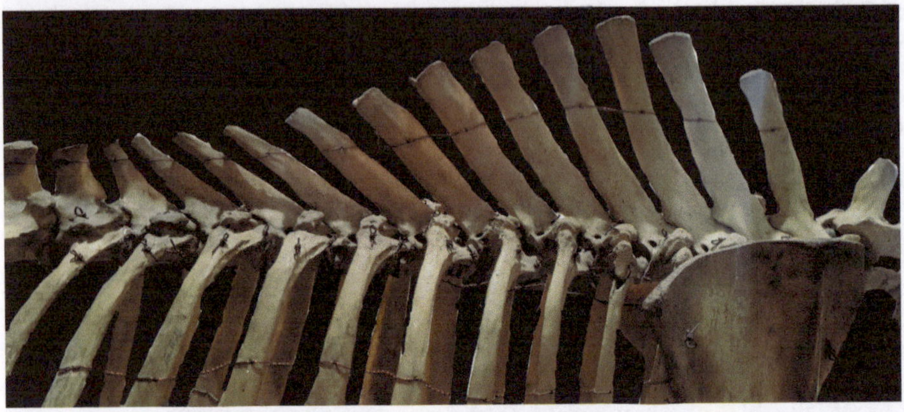

Fig. 2.36: Photograph Showing the Thoracic Vertebrae of Ox

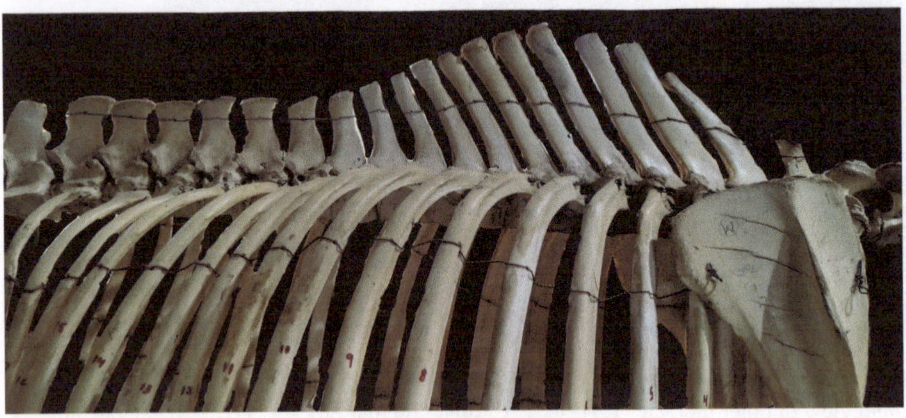

Fig. 2.37: Photograph Showing the Thoracic Vertebrae of Horse

c. Comparative Study of Lumbar Vertebrae

Lumbar Vertebrae of Ox

Lumbar vertebrae are six in number and form the roof of the abdomen. They are recognized by the great size and form of the transverse process (Fig. 2.37). The body is much compressed from above downward, constricted in the middle. Body of last lumbar is compressed dorsoventrally and articulate with cranial end of sacrum. Neural ring is roughly triangular. Ventral crest is present. The dorsal Spinous process is in the form of broad, flattened quadrilateral plate. The height of the dorsal spinous process is similar to the last thoracic vertebra. Their summit is moderately thickened. The spinous process of the first lumbar vertebra is the narrowest of all. The last one is being smallest. The convexity and concavity of cranial and caudal extremity of the body is less (Fig. 2.37). The body of the 4th and 5th lumbar vertebrae is relatively more longer than others. The breadth and height of the arch is gradually increase from 3rd to 6th lumbar. The caudal notches are much deeper. The lateral vertebral foramina are often double in the cranial end of the series. The Articular processes are well developed and large. The cranial articular processes are fused with the very prominent mammillary processes and form mamilloarticular process. Cranial articular processes are curved inward and present dorsally a deep concave surface for articulation with the caudal pair of articular processes of the proceeding vertebra (Fig. 2.37). The caudal articular processes are convex and pointed and have ventro-laterally strong convex articular surfaces which fit into the concave surfaces of the cranial pair of the succeeding vertebra. Transverse processes are in the form long bony plates, flattened dorso-ventrally and inclined cranially. They have thin and irregular borders, and often bear projections of variable size and form. Transverse processes of first lumbar are shortest and the length increases to the fifth, the last being considerably shorter and narrower. *Muscle intertransversales lumborum* are occupying the spaces between the lumbar transverse processes. The *Multifidus Dorsi Muscle* originates from the articular processes of the lumbar and the transverse processes of the thoracic vertebrae and insert on the caudal border of the spinous processes of the first four lumbar and all of the thoracic vertebrae. The *Muscle longissimus costarum* took origin from the cranial borders of the ribs except the first five, and the ends of the lumbar transverse processes.

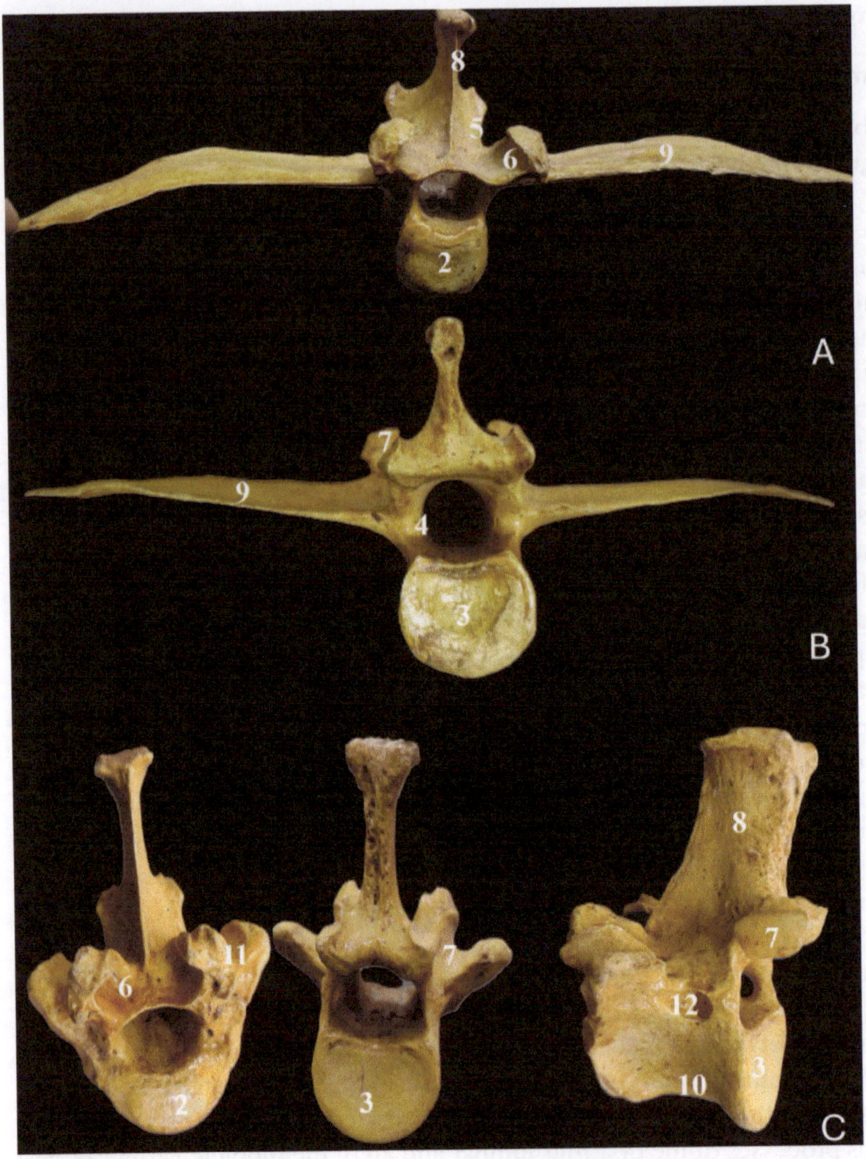

Fig. 2.38: Photograph showing the Lumbar vertebrae of Ox A. cranial view, B. Caudal view, C. 6th lumbar vertebrae (cranial, caudal and lateral view)

1. Body
2. Cranialextremityof the body
3. Caudalextremity of the body
4. Pedicle of arch
5. Lamina of arch
6. Cranial articular process
7. Caudal articular process
8. Dorsal spinous process
9. Transverse process
10. Ventral spinal process
11. Mamillary process
12. Lateral vertebral foramen

Table 2.23: Comparative table on lumbar vertebrae of ox, horse, dog and pig

Characters	Ox	Horse	Dog	Pig
Number	6	6	7	6-7
Body	Much compressed from above downward, constricted in the middle and bears ventral crest	Shorter than ox ventral crest is distinct on the first three and less distinct in the fourth and absent on the last two (5th and 6th)	Flattened dorsoventrally, increase in width from first to last. The length increases to the sixth	Flattened dorsoventrally
Transverse process	Horizontally oriented. Inter transverse articulation is absent. 1st has smallest and 3rd or 4th has longest transverse process	Horizontally oriented, 5th and 6th and sometimes 4th articulate by corresponding convex facet cranially and concave facet caudally. intervertebral foramen at this fused region ha dorsal and ventral opening	Directed forward and downward Longest in 5/6, cranio-ventral inclination. Inter transverse articulation is absent	Inclined cranio-ventrally. Inter transverse articulation is absent
Articular process	Cranial articular processes- are curved inward and present dorsally a deep concave surface. Caudal processes are convex and pointed and have ventro-laterally strong convex articular surfaces	Same as in ox	Flat sagittally directed surface	Same as in ox
Dorsal Spinous process	Broad flattened plates and are of the same height as the last thoracic vertebra. Their summit is moderately thickened. The spinous process of the first lumbar vertebra is the narrowest of all. The last one being the smallest Inclined caudally	Higher than in ox, Flat, curved slightly cranially up to 4th , free end expands to form ridge	Broad ventrally, narrow dorsally. Inclined cranially except last height diminishes caudal to the fourth	Broad with sharp border, directed cranially
Mamillary process	Fused with cranial articular process to form combined mamilloarticular process	Plate like	Same as in ox	Present except the first two, Project lateral and caudal
Accessory process	Not present	Not present	Present between transverse and caudal articular process,. It project caudally upto 5th vertebrae	Not present

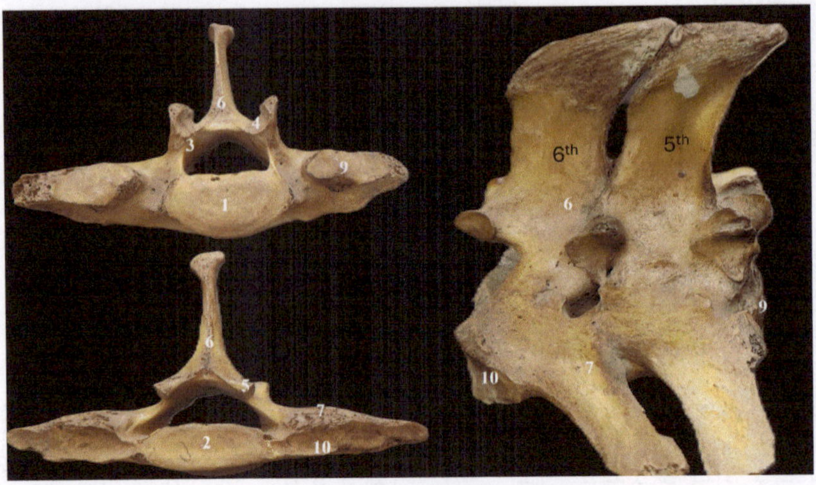

Fig. 2.39: Photograph showing the Lumbar vertebrae of horse A. 6th lumbar (cranial view), B. 6th lumbar (caudal view), C. 5th and 6th articulated lumbar (lateral view)

1. Cranial extremity of body
2. Caudal extremity of body
3. Arch
4. Cranial articular process
5. Caudal articular process
6. Dorsal spinous process
7. Transverse process
8. Lateral vertebral foramen
9. Facet present on the cranial surface of transverse process
10. Facet present on the caudal surface of transverse process

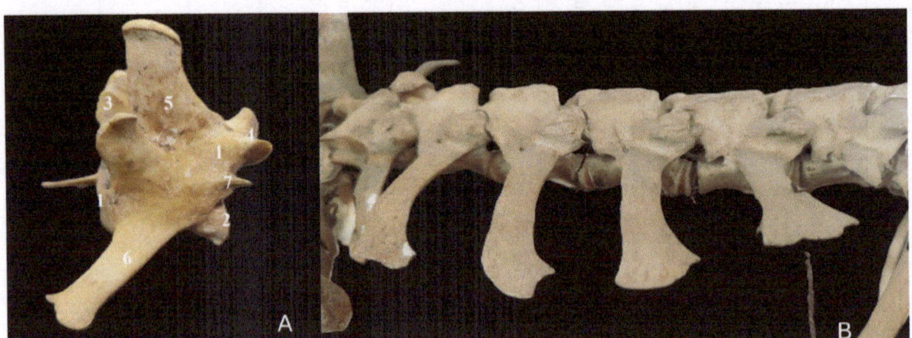

Fig. 2.40: Photograph showing the Lateral view of Lumbar vertebra of Dog (A) and Pig (B) 1. Cranial extremity of the body, 2. Caudalextremity of the body, 3. Cranial articular process, 4. Caudal articular process, 5. Dorsal spinous process, 6. Transverse process, 7. Accessary process

d. Comparative Study of Sacral Vertebrae

Sacral Vertebrae of Ox

The sacrum is formed by the fusion of five sacral vertebrae which involves fusion of the dorsal spinous processes and articular processes. It is roughly triangular in outline and form the roof of the pelvic cavity along with first three coccygeal vertebrae. Its long axis is strongly curved. It articulates cranially with last lumbar vertebra, caudally with first coccygeal vertebra and craniolaterally with ilium. It presents two surfaces, two borders, a base and an apex.

The dorsal surface presents centrally the median sacral crest which is formed by fusion of dorsal spinous process. It is rough for attachment of ligaments and muscles. A groove is present on either side of the crest which is bounded laterally by four pairs of the lateral/ dorsal sacral foramina through which the dorsal branches of sacral nerve emerge. Cranial two pairs of foramina are situated on the medial side and caudal are situated on lateral side of lateral sacral crest. Lateral sacral crests are low in height situated parallel to median sacral crest on either side and are formed by fusion of articular processes. The pelvic surface is wide infront and somewhat narrow behind and concave in both directions. It is marked by four more or less distinct transverse lines, which indicates the demarcation of the bodies of the vertebrae. A groove is marked centrally which designate the passage of middle sacral artery. Four pairs of oval, ventral sacral foramina are present at the end of these lines for transmission the ventral division of the sacral nerves. These foramina are larger than the dorsal sacral foramina and reduce in size from first to last. The dorsal and ventral foramina communicate with the sacral canal and are together equivalent to the usual lateral vertebral foramina. Two lateral borders are thin, sharp, and irregular (Fig. 2.40).

The base is wide and directed forward. It is formed by cranial end of the body, cranial articular process and neural ring. Body is dorsoventrally flat. The body of the first sacral segment is wide transversely, dorso-ventrally flattened, and has a rounded surface which articulates with the body of the last lumbar vertebra. The ventral margin of body of 1st sacral vertebrae projects slightly downward, forming the promontory. At the entrance of neural canal, a pair of articular processes flanked above the body, which project upward and forward from the arch, and overhang the body. They have concave articular surfaces medially.

The lateral parts of the base form wings, which curves downward and forward and are quadrangular, short masses, compressed from before backward. The wings have an extensive cranial non-articular surface which is concave from side to side. The caudal surface is rough and at its lower part there is triangular area for articulation with Ilium. Dorsal margin of the wing is roughened for attachment of sacro-iliac ligament. Apex and is the caudal aspect of the last sacral vertebra. It is wide transversally and has circular body. The transverse processes projecting outwards on each side from the body. Neural ring is triangular and leads to caudal end of neural canal.

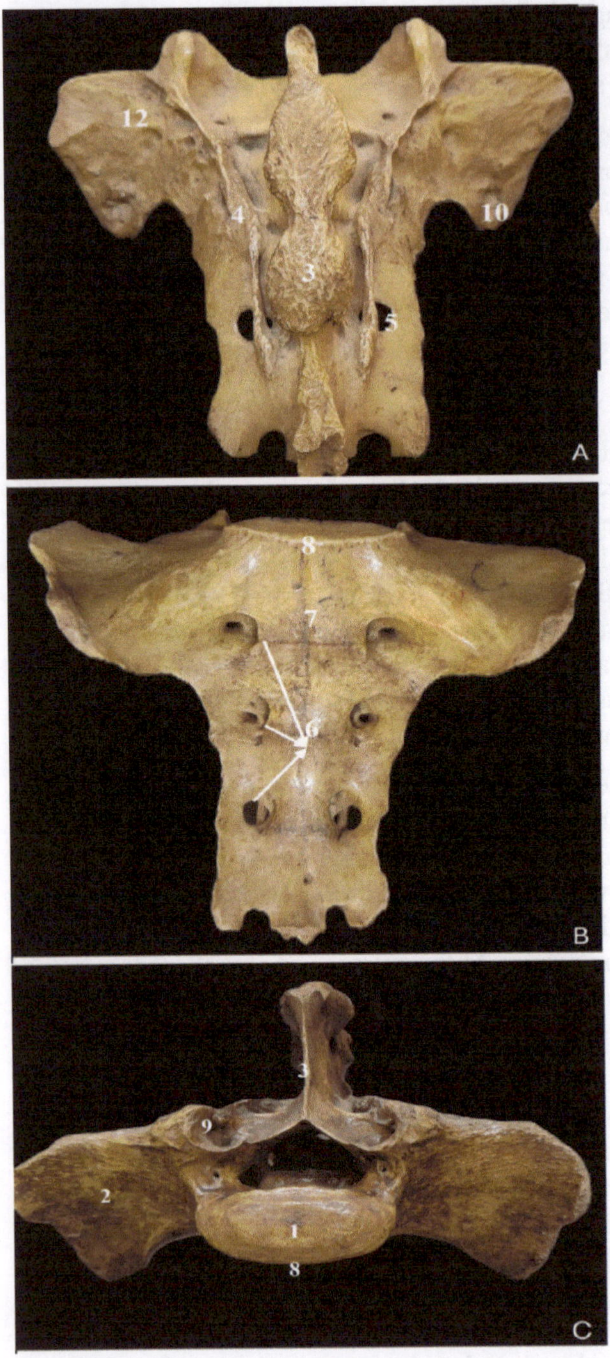

1. Cranial extremity of the body,
2. Wing of sacrum,
3. Median sacral crest,
4. lateral sacral crest,
5. Dorsal sacral foramina,
6. ventral sacral foramina,
7. Pelvic surface of sacrum,
8. Sacral promontory,
9. Cranial articular process,
10. Area for articulation with ilium,

Fig. 2.41: Photograph showing the Sacrum of Ox A. Dorsal view, B. Ventral view and C. Cranial view

Table 2.24: Comparative table on Sacral Vertebrae of Ox, Horse, Dog and Pig

Characters	Ox	Horse	Dog	Pig
Number	Consist of 5 sacral segments. base and apex	Same as in ox	Consist of 3 sacral segments.	consist of 4 sacral segments.
Shape	Roughly Triangular	Triangular	Quadrilateral	Triangular
Fusion completed	3-4 years	4-5 years	1.5 years	1.5 years
Body	It is longer than the horse	It is shorter than the ox	It is short, wide and quadrangular in shape	Pelvic surface is less curved as compared to ox
Wing	Originates from 1st sacral vertebrae, very broad. Quadri-angular, curve downwards and forwards. An extensive non articular cranial surface. The caudal surface is rough and at its lower part there is triangular area for articulation with ilium.	Prismatic, a large oval convex facet infront to articulate with the facet present on the transverse process of the last lumbar, vertebra. Caudally an elongated oval area (auricular facet) for articulation with the ilium.	2nd sacral vertebrae also participate in wing formation, Wings are prismatic and very high and bear laterally, auricular surface on its lower part. Anterior processes are large and have extensive, slightly concave facets.	Consists of transverse process of 1st and 2nd cranial articular processes are very large
Dorsal surface	Dorsal surface presents spines fuse to form median sacral crest. Articular processes fuse to form lateral sacral crest and 4 pairs of dorsal sacral foramina.	Dorsal surface presents 5 sacral spines which have tuberous summits except 1st. The fusion of the spines is not so complete as in ox.	Spines are fused to form median crest which is notched. On either side 2 tubercles are the vestiges of the fused articular processes.	Spinous processes are replaced by indistinct crest.
Ventral surface	Deeply arched, concave in both directions presents four transverse groove and four pairs of foramina. presents shallow median groove for middle sacral artery.	Not so deeply arched as in ox, lodges four transverse lines and having four pairs of foramina.	Deeply concave and has two pairs of foramina.	Middle of the dorsal surface is flattened, smooth and presents opening.

Characters	Ox	Horse	Dog	Pig
Lateral sacral crest	Indistinct/ insignificant	Distinct	Indistinct	Distinct
Notches	No notches on either side of the body of 1st sacral segment.	Smooth notches are present on either side of the body of 1st sacral segment.	Sacral canal is compressed dorso-ventrally	Transverse lines are distinct
Articular surface	Articular surfaces are concave and semicylindrical in curvature medially and processes are large and widely separated.	Articular surface directed dorso-laterally which is slightly concave in its length. It is rough, irregular and articulates with ilium. Portion lateral to the foramina is formed by the fusion of the transverse processes.	Transverse processes of last vertebrae project backward.	Same as in ox
Dorsal sacral foramina	Four pairs, Large between last two and others are smaller	Four pairs, Smaller than ventral sacral foramina	Two pairs, Similar in size of ventral sacral foramina	Three pairs, and Narrow
Ventral / pelvic sacral foramina	Extremely large and oval four pairs in number	Larger than dorsal sacral foramina, diminish in size from 1st to last, four pairs in number	Similar in size of dorsal sacral foramina, two pairs in number	Extremely wide, three pairs in number

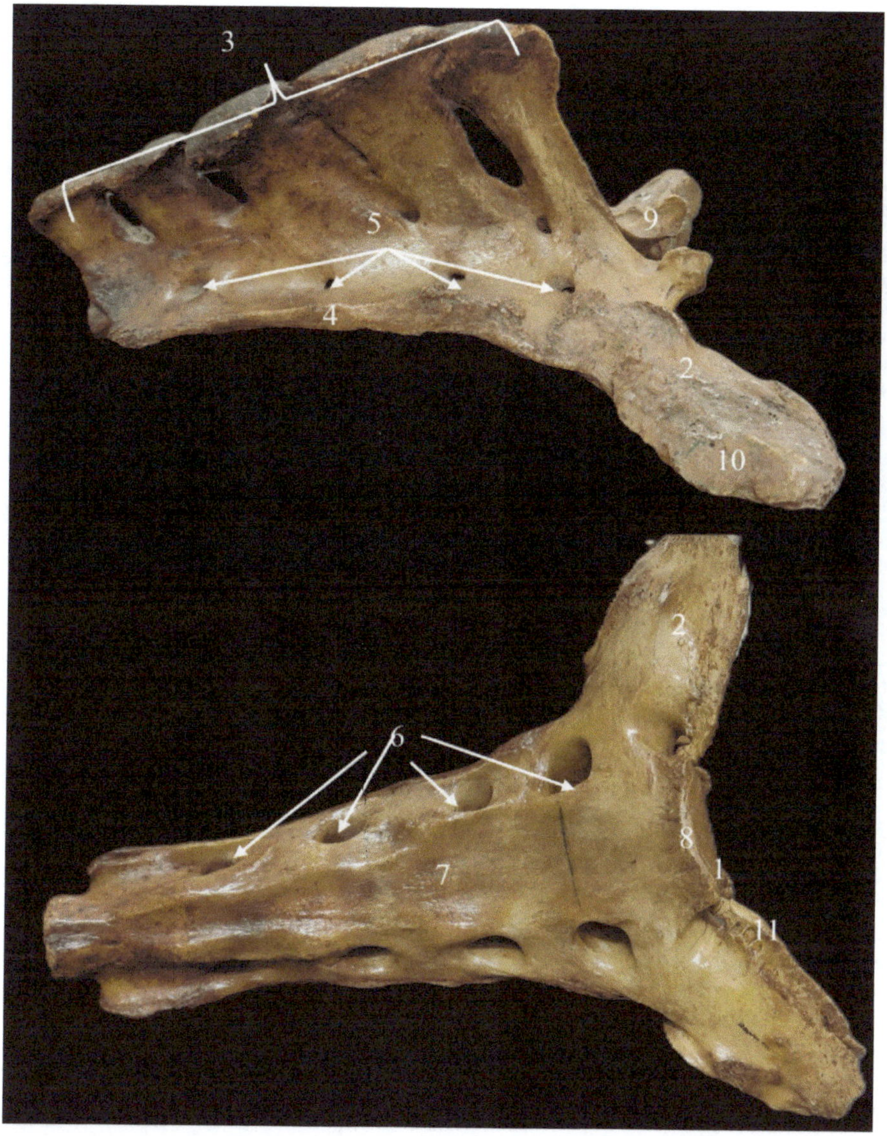

Fig. 2.42: Photograph showing the Sacrum of Horse A. Dorsal view and B. Ventral view

1. Cranial extremity of the body
2. Wing of sacrum
3. Dorsal sacral crest
4. Lateral sacral crest
5. Dorsal sacral foramina
6. Ventral sacral foramina
7. Pelvic surface of sacrum
8. Sacral promontory
9. Cranial articular process
10. Facet for articulation with ilium
11. Facet for articulation with 6th lumbar vertebra

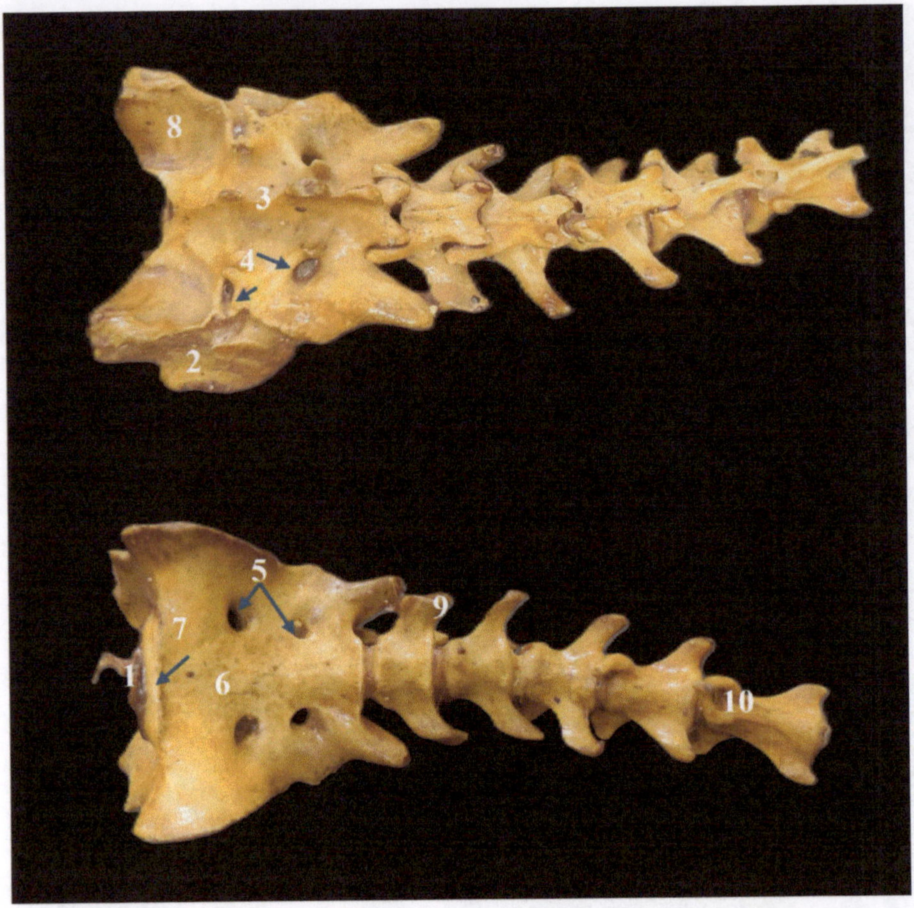

Fig. 2.43: Photograph showing the Sacrum and coccygeal vertebrae of Dog A. Dorsal view, B. Ventral view 1. Cranial extremity of the body, 2. Wing of sacrum, 3. Dorsal sacral tubercle, 4. Dorsal sacral foramina, 5. Ventral sacral foramina, 6. Pelvic surface of sacrum, 7. Sacral promontory, 8. Cranial articular process, 9. Transverse process of coccygeal vertebrae, 10. Hemal arch

e. Comparative Study of Coccygeal Vertebrae

Coccygeal Vertebrae of Ox

Coccygeal vertebrae are in the range of 18 to 20 which are incomplete vertebrae and form the skeleton of tail. First five vertebrae have neural ring, body and processes. Their bodies are flattened dorso-ventrally, constricted in the middle, and at the ends have convex, elliptical, articular surfaces. The arch is low and triangular. The transverse processes are relatively large in the cranial vertebrae. The articular processes are ill developed and do not carry articular facets. The size of the body gradually reduces and last few presents only rod-shaped bodies with convex ends. Ventrally they form median groove, the *sulucus vasculosus* for middle coccygeal artery. The arch disappears after sixth backwards and the rudimentary transverse process disappears at the ninth or tenth coccygeal vertebra. *Dorsal and lateral sacrococcygeus muscle* originates from dorsal to the transverse processes. *Ventral sacrococcygeus muscles* originate ventral to the transverse process of coccygeal vertebrae. It originate from the transverse process of the fifth sacral vertebra and its insertion is to the transverse processes of all of the coccygeal vertebrae. The *Muscle intertransversalis caudae* are located between the transverse processes. The *Muscle coccygeus* get Insertion to the transverse processes of the second and third coccygeal vertebrae.

Table 2.25: Comparative table on Coccygeal Vertebrae of Ox, Horse, Dog and Pig

Characters	Ox	Horse	Dog	Pig
Number	18 to 20 in number.	15 to 21 (18) in number.	20-23 in number	20 to 23 in number.
Body	Flattened dorso-ventrally, constricted in the middle. Both cranial and caudal surface are convex elliptical, articular surfaces	More flattened body than ox	Gradually increase in length from 5-10, then progressively shorten	Same as in ox
Arch	low and triangular by 7 disappear completely	From 3rd coccygeal it is incomplete	Incomplete from 8th	First 5 have complete arch
Vertebral canal	Progressively disappears far as the 5th, open dorsally in 6th in the form of groove	Open dorsally from 3rd	Open from 8th in the form of groove	Open from 5th or 6th
Dorsal Spinous processes	Largest in 1st, gradually diminishes upto 5 and then disappears	Short and have double summit. 2nd is bifurcate	Form of low ridge diminishes from 8th onwards	Ridge like, First 3 or 4 vertebrae are specially characterized by the presence of functional articular processes.

Characters	Ox	Horse	Dog	Pig
Transverse process	Long and broad in 1st, gradually smaller in size upto 9th and then disappear	In the form of small bony ridge upto 6th and 7th it is absent	Diminishes from 5th vertebrae onwards	1st three in the form of broad plate which gradually become smaller, and after 9/10 represented by nodule
Mamilloarticular process	Absent	Absent	Present being last to disappear among all processes	Absent
Haemal process	Paramedian process on ventral aspect of body (cranially) of 1st to 8th vertebrae	Not present	Same as in ox, present in 5th to 15th vertebrae	Not present
Ventral arches / haemal arch	Formed by haemal process from 2nd -3rd	Not formed	Formed by haemal process from 3rd -8th coccygeal vertebrae. These bones are known as "Chevron bones"	Not formed

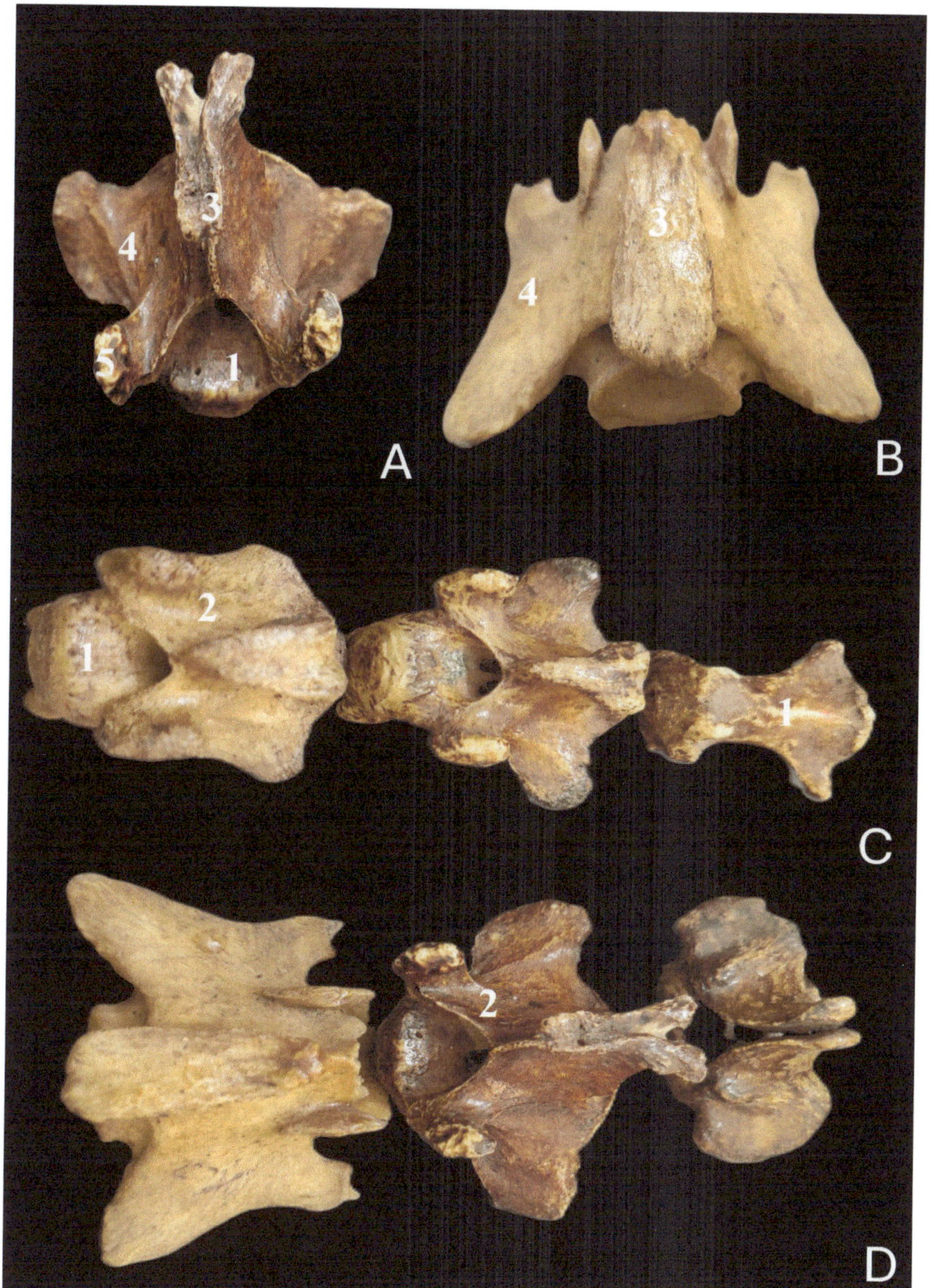

Fig. 2.44: Photograph showing the coccygeal vertebrae of Ox (A and C), Horse (B and D), 1. Body, 2. Arch, 3. Dorsal spinous process, 4. Transverse process, 5. Cranial articular process

C. Comparative Study of Ribs

These are elongated curved bones arranged one after another on either side of vertebral column and form the skeleton of the lateral thoracic walls. Each of them are arranged serially in pairs which correspond in number to the thoracic vertebrae. Each of them articulates above with two vertebrae (last cervical and all thoracic vertebrae). A rib consists of the bony part and the cartilaginous part. The cartilaginous part of a rib is termed as "the costal cartilage". The space between two ribs is known as the intercostal space. Ribs from different parts of the series vary greatly in length, curvature and other characters.

Kinds of Ribs: Ribs are of three kinds:

(1) The Sternal Rib or True Rib: The bony part of a rib articulates above with the vertebrae and cartilaginous part (ie., the costal cartilage) articulates below with sternum.

(2) Asternal Rib or False Rib: is that whose bony part articulates above with the vertebrae but whose costal cartilage does not articulate with the sternum, but is overlapped and is attached to the preceding or other costal cartilage by elastic tissue to assist information of the costal arch.

(3) Floating Rib: is that rib whose bony part articulates above with the vertebrae and below its cartilage is free and is not attached to adjacent costal cartilages.

Table 2.26: Showing the number of ribs in different domestic animals.

Species	Sternal rib	Asternal rib	Floating ribs	Total
Ox	8	5	0	13
Horse	8	10	0	18
Dog	9	4	Last rib (13th)	13
Pig	7	7	0	14-15
Fowl				7

Rib of Ox

In ox, there are 13 pairs of ribs, out of which first eight pairs are sternal and remaining five pairs are asternal. The shape and size of the ribs varies at different levels.

A rib comprised of a shaft or body and two extremities. A typical rib has a shaft, a head at the proximal end and a costal cartilage at the distal end (Fig. 2.44). The shaft is curved and presents two surfaces (lateral and medial) and two borders (cranial and caudal). The curvature is more pronounced in the dorsal third. The lateral surface is convex in length and transversally and has a wide groove at upper end. The medial surface is concave in its length. Cranial border starts from head and is thick and concave. Caudal border starts from tubercle and is also convex. It has a costal groove for intercostal vessels and nerves. The borders give attachment to the *external and internal intercostals muscles*. The upper extremity or the vertebral extremity consists of the head, a tubercle and a neck (Fig. 2.44 and 2.45). The head presents two convex articular facets separated by a non-articular groove, *conjugal groove* which gives attachment to the conjugal ligament. Both these facets articulate with the cavity formed by the costal facets on the bodies of two adjacent thoracic vertebrae and the intervertebral fibro cartilage. The neck is the constriction below the head, and it roughened above and infront for ligamentous attachment. The tubercle is placed above and behind the junction of neck and body. It has an oval facet for articulation with the facet present on the transverse process of the corresponding thoracic vertebra. The distal end/ lower extremity or sternal extremity is expanded and articulate with costal cartilage. In last rib the facets as the tubercle and those on the head are confluent.

The eight, ninth and tenth ribs are the longest and the widest among all ribs. The length and breadth gradually reduce infront and behind of these ribs. The size of the head and tubercle diminishes from the first to last. The superficial part of the muscle *Dorsal scalenous* inserts on the fourth rib and deep part *of dorsal scalenus and ventral scalenus* muscles ends on the anterior border of first rib. The *Muscle longissimus costarum* originates from the anterior borders of the ribs except the first five, and the ends of the lumbar transverse processes and attaches on the posterior border of all the ribs. The *Muscle levatores costarum* is located at the upper part of the intercostal space and are blended with the *external inter costal muscles* at their insertion on the anterior border of rib. The *Muscle intercostalis externi (external intercostal muscle)* do not extend into the interchondral spaces. It originates from the posterior border of the rib and insert on the anterior border of the succeeding rib. Similarly, the *Muscle intercostalis interni* (*internal intercostal muscle) are* thinner than the external muscles and extend into the interchrondral spaces. These muscles arise from the cranial border of the ribs except the first rib and ends on the posterior border of the preceding rib. The *Muscle retractor costae* is a triangular muscle which lies in the angle formed by the last rib and the ends of the lumbar transverse processes. Ventral half of the first rib gives origin to the *Muscle rectus thoracis* which inserts at the third of fourth costal cartilage.

Table 2.27: Showing the difference between 1st rib and typical rib of ox

Features	Typical rib	First rib of ox
Body	Elongated, wide, flattened and curved	Shortest and least curved
Conjugal groove	Present	Absent
Articular facets on the head	Facets are separated by conjugal groove	Cranially small and caudally large facet
Neck	Distinct	Thick and very short
Tubercle	Oval facet	Largest among all rib and convex in length and extensive facet
Sternal end	Elongated	Larger, thick and wide than other ribs
Sub costal groove	Caudal side of medial surface	Absent

Ribs of Horse

1. There are eighteen pairs of ribs - eight pairs of them are sternal and the rest (9th to 18th) are asternal.
2. In general, they are shorter, narrower, and more curved in form than the ribs of ox.
3. The length of the ribs increases from the first to the tenth or the eleventh and rib and then diminishes.
4. The width of the ribs increases from the first to the sixth rib and then diminishes.
5. The cranial border is thin and sharp from second to eight ribs behind which this border becomes thick and rounded.
6. The curvature gradually increased from second to the seventh riband than from seventh to sixteenth and eighteenth the curvature again decreases noticeably.
7. The neck of the rib is thin and remain same and is absent from the last two or three ribs.
8. The lower ends of the ribs are excavated for receiving the costal cartilages.

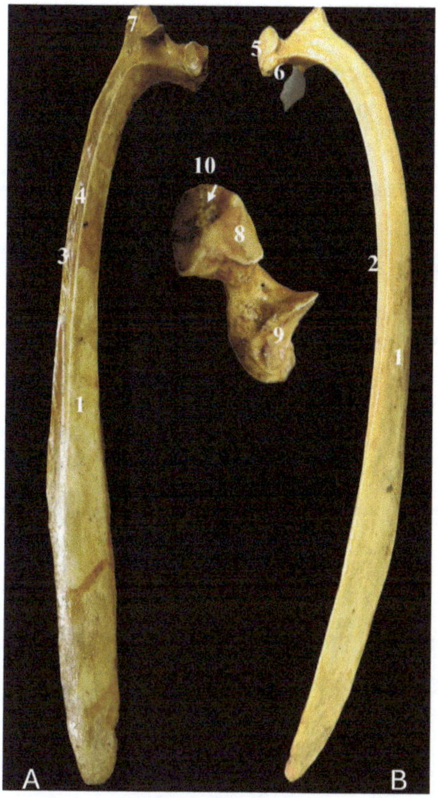

Fig. 2.44: Photograph showing the Rib of Ox
A. Medial surface, B. Lateral surface: 1. Shaft, 2. Cranial border, 3. Caudal border, 4. Costal Groove, 5. Head, 6. Neck, 7. Tubercle, 8. Facets of head, 9. Facet of tubercle, 10. Conjugal groove

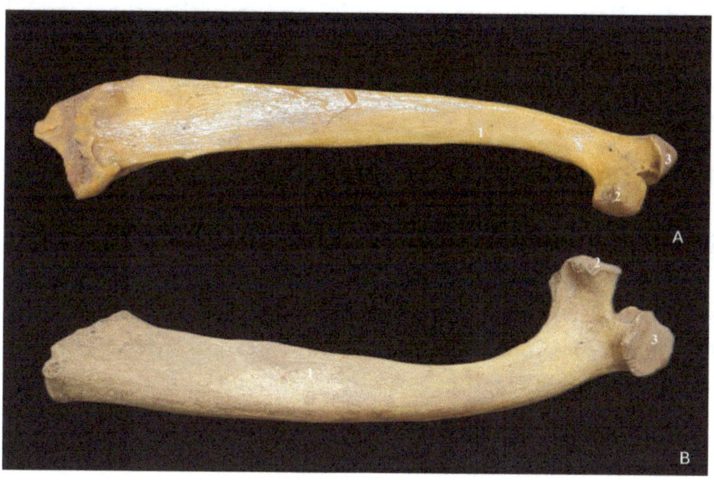

Fig. 2.45: First rib of A. Ox, B. Horse: 1. Shaft, 2. Fead, 3. Tubercle

Ribs of Dog

1. There are thirteen pairs of ribs. First nine pairs are sternal, next three pairs are always asternal where, the last pair is either asternal or floating in character.
2. The width in the lower part of the rib increases in the first eight or nine ribs.
3. They are narrower, relatively thicker, and more curved than in other domestic animals.
4. Each of the last three ribs articulates with only one vertebra corresponding to the more caudal posterior of the two vertebrae of earlier joints and the facet on the head is therefore undivided in these three ribs.

Ribs of Pig

1. There are 14 or 15 pairs of ribs, out of which 7 are sternal and 7 or 8 are asternal.
2. They are strongly curved with a distinct angle.
3. The first rib is prismatic, has a large sternal end and a very short cartilage.
4. The tubercle of the rib fuse with the head in last five or six ribs.
5. 2nd to 5th rib form diarthrodial joint with its cartilage.

D. Comparative Study of Sternum

Sternum of Ox

Sternum is an osteocartilagenous structure placed in the middle of the floor of the thoracic cavity and articulates with the cartilages of the sternal ribs laterally. It is inclined obliquely so that the cranial end is higher than caudal end. In consists of seven sterna brae (bony segment). For description it is divided into three parts viz. Manubrium sterni, Meso sternum and Xiphoid cartilage (Fig. 2.46).

The cranial extremity or pre-sternum or the manubrium sterni is formed by the first sternebra. It is wedge shaped and laterally compressed (Fig. 2.46). Its base forms a diarthrodial joint with the body of the bone, and laterally it bears extensive facets for articulation with the first pair of costal cartilage. *Muscle sternothyrohyoideus* originates from the cranial surface of the manubrium sterni. The *Muscle sternocephalicus* extends from the sternum to the head on the ventro lateral part of the neck. Both parts of *sternocephalicus* have their caudal attachment to the manubrium sterni and the first costal cartilage

The Meso-Sternum or Body is formed by second to seventh (six) sternabrae. The width of the body increases from 2nd to 6th and becomes much narrower behind the last pair of costal facets (Fig. 2.46). The dorsal surface is extensive and is strongly convex infront but more or less concave behind the second sternebra. Its ventral surface is concave longitudinally and in the form of very narrow isoscale triangular with the apex infront. Both the surfaces are marked by transverse lines indicate the lines of fusion of the sternebrae. But last two sternebrae fuse completely to form a single mass and there is no line of demarcation. Lateral borders are thick. Seven articular depressions known as the costal cavities are present on the lateral surface. These cavities are situated in series at the inter-sternebral junctions. Costal cartilage of 8th pair of sternebrae joins the sternum between 6th and 7th sternebrae along with costal cartilages of 7th pair (Fig. 2.46).

The caudal extremity is formed by xiphoid cartilage which is a thin plate attached to last sternebra. Its dorsal surface is concave, wide and covered by superior ligament and transverses thoracis muscle and gives attachment to diaphragm. The ventral surface is convex and furnishes attachment to the *transversus abdominis muscle and the lines alba*. The free margin is thin. The *Muscle transversus thoracic* lies on the dorsal surface of the sternum and the medial surface of the cartilages of the second to the sixth ribs. The sternal part of the diaphragm is attached to the abdominal surface of the xiphoid cartilage (Fig. 2.46).

Table 2.28: Comparative table on Sternum of Ox, Horse, Dog and Pig

Characters	**Ox**	**Horse**	**Dog**	**Pig**
No. of Sternebrae	7	7	8	6
Shape	Dorso-ventrally flattened	Canoe or boat shaped	Ventrally convex bony rod	Same as in ox
Manubrium sterni	Dorsally curved	Long, dorsally convex and laterally compressed,	Sharp blunt cylindrical in shape, Oval in cross section	Long, blunt cone shaped cartilage
Body	Wide and Dorso-ventrally flattened	ventral border forms the prominent keel-like crest of the sternum	rectangular	Triangular
Xiphoid cartilage	Broad	Broad plate	Long dorso-ventrally flattened	Short and narrow
Cariniform cartilage	Absent	Laterally compressed cartilage present cranially, with a deep notch dorsally for articulation with first two costal cartilage	Short conical cartilage at the cranial end of manubrium	Absent

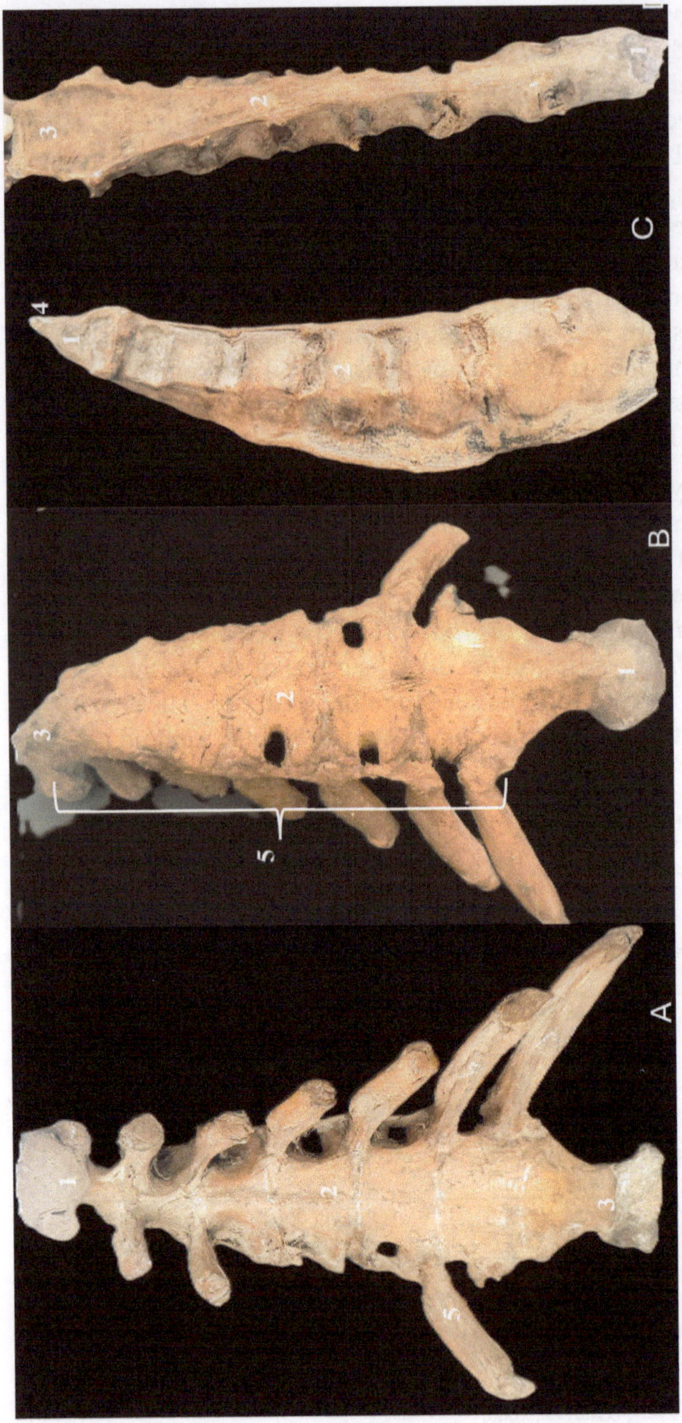

Fig. 2.46: Photograph showing the Sternum of Ox (A. Dorsal view, B. Ventral view) and Horse (C. Lateral view, D. Dorsal view): 1. Manubrium sterni, 2. Mesosternum, 3. Xiphisternum, 4. Cariniform cartilage, 5. Costal cartilage

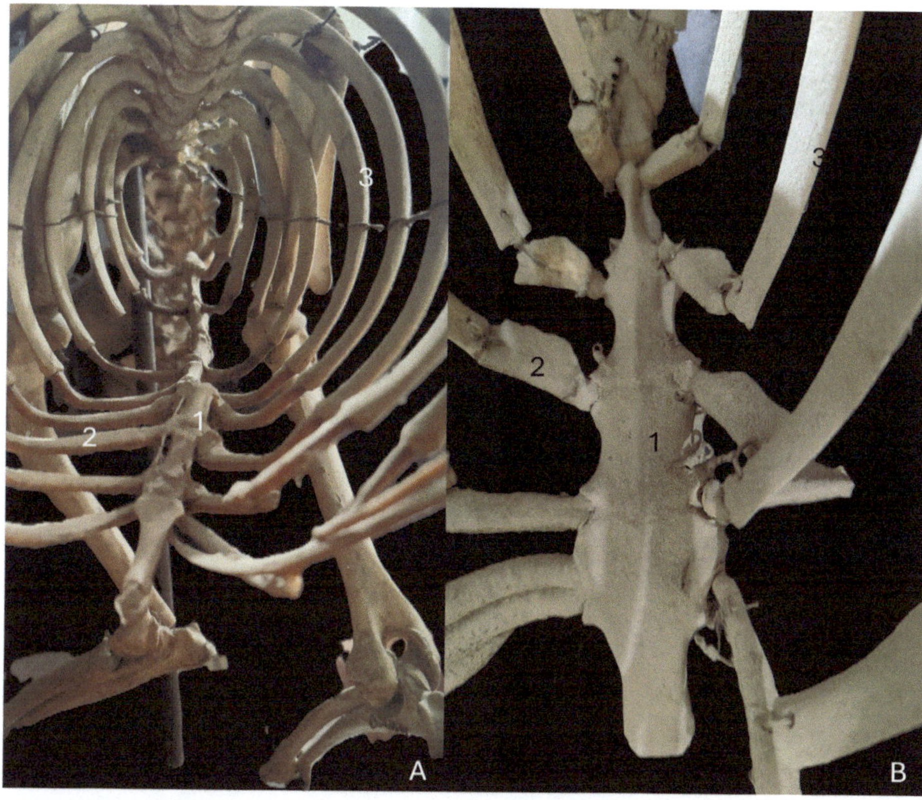

Fig. 2.47: Photograph showing Ribs and sternum of Dog (A) and Pig (B) 1. Sternum, 2. Costal Cartilage, 3. Ribs

Costal Cartilage of Ox

These are bars of hyaline cartilage which continue the ribs. There are thirteen pairs, first eight, are sternal and rest asternal. The first pair of costal cartilage is very short and articulate by their medial surfaces with the sternum but not with each other. The dorsal end is very wide and thick. Their sternal end is small. The cartilages of the other sternal ribs increase progressively in length and become more rounded. The sternal end is enlarged and has an elliptical convex facet for articulation with the sternum. The cartilage of the asternal ribs is long slender, and pointed. The cartilages of the asternal ribs overlap and are attached to each other by elastic tissue, forming the costal arch. Except in case of the first, the cartilage does not continue the direction of the rib but forms with the latter an angle which is open infront, and increases from second to last.

Costal cartilage of horse: The first pair of costal cartilage articulates with each other as well as with the sternum.

Costal cartilage of dog: They are long and curve ventrally. The length and curvature of the first pair of costal cartilages are strikingly special.

Costal cartilage of pig: wide and plate like. 2-5th form diarthrodial joint with ribs.

3

Appendicular Skeleton

A. Comparative Study of the Skeleton of Pectoral Limb

The fore limb of ox and other quadruped have four recognized regions *viz.* shoulder (shoulder girdle), the arm (brachium); the fore arm (antibrachium); and the fore foot (manus).

The skeleton of the shoulder girdle, when fully developed consists of three bones – scapula (shoulder blade), the coracoids, and the clavicle or collar bone. However in horse and ox the skeleton of this region is made up of only one bone, the Scapula. The clavicle is absent. In case of dog shoulder girdle has one or two bones - scapula and clavicle. The clavicle is small, thin, irregularly triangular cartilaginous or bony plate which is embedded in the brachiocephalicus muscle cranial to the shoulder joint. The clavicle does not form any articulation with rest of the other skeleton of the body. In fowl all the three bones scapula, coracoids and clavicle are present.

The skeleton of the arm is invariably formed by a single bone, - the humerus. The fore arm, consist of two bones the radius and ulna. The manus which is the homologues of the hand in man consist of three subdivisions viz., the carpus, metacarpus and digit/ digits.

a. Comparative Study of Scapula

Scapula of Ox

Scapula forms the shoulder region of the body. It is a flat bone in which length is more than width and thickness. The bone is situated obliquely downward and forward or distally cranially from fourth thoracic spine to sternal end of first rib. For description it presents two surfaces *viz.* lateral and medial surface, two borders *viz.* cranial and caudal border and three angles *viz.* cranial, caudal and glenoid angle (Fig. 3.1).

The Lateral surface is divided into two fossae by the spine of the scapula which extends from the vertebral border to the neck of the bone where it becomes more prominent, and is prolonged by a pointed projection called acromion process from which a part of *deltoid muscle* arises. The spine is sinuous (wavy), bent backward in its middle (attachment of *trapezius muscle)*, forward below, and its main part is subcutaneous. The supraspinous fossa is situated infront of the spine. It is narrow, and does not extend to the lower part of the bone. It is smooth and is occupied by the *supraspinatus* muscle. Infraspinous fossa is situated behind the spine and is very wide extends right down to the lower part of the bone and occupied by *infraspinatus muscle*. The lower part of infraspinous fossa is narrow and marked with several rough lines for origin to the *teres minor muscle*. The area of infraspinous fossa is about three times more than the area of supraspinous fossa.

Medial (costal) surface is hollowed in its length by the subscapular fossa for lodgement *of subscapularis muscle*. The fossa separates two rough areas above, the cranial one is quadrilateral (for the attachment of *Serratus cervices muscle*) while the caudal is oblique and slightly curved (for the attachment of *Serratus thoracis muscle*) (Fig. 3.1) .

The cranial (Anterior) border is convex and rough above, concave and smooth below. Caudal (posterior) border is slightly concave, thick and rough in its upper third, thin in its middle and again thick below. *Deltoideus, teres major and long head of triceps muscles* attach on this border. The nutrient foramen is usually situated in the lower third of this border. Dorsal (Vertebral) border carries un-ossified part of the foetal scapula-termed the scapular cartilage. This edge is thick and pitted into which cartilage is fitted. The medial surface of scapular cartilage gives attachment to *Rhomboideus muscle* (Fig. 3.1).

Cranial (Anterior /Cervical) angle is located at the junction of the cranial and vertebral borders and lies opposite the second thoracic spine. It is relatively thin and is about right angle. Caudal (Posterior) angle is thick and lies against the sixth rib (and its position can be determined readily in the living animal) (Fig. 3.1).

Glenoid (Ventral/Articular) Angle is joined to the body of the bone by the neck of scapula. It bears circular glenoid cavity for articulation with the head of the humerus. The rim of the cavity presents the glenoid notch on cranio-lateral aspect. Infront of glenoid cavity a supra glenoid tubercle called tuber scapulae is present which furnishes origin to *Biceps brachii muscle*. A small coracoid process is located on medial aspect of cavity which gives origin to *coracobrachialis muscle* and lies close to the glenoid cavity.

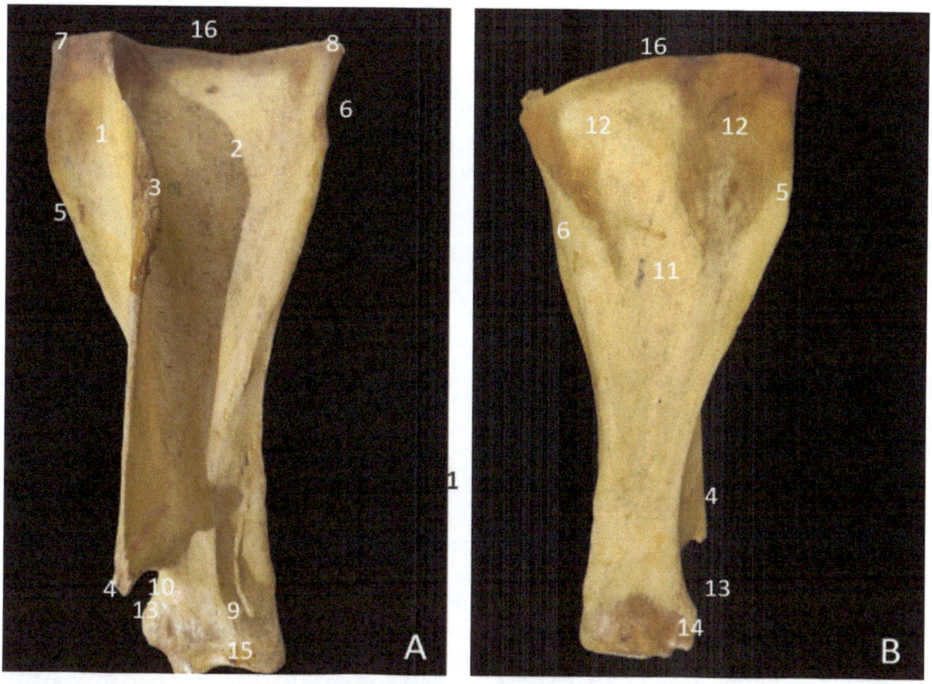

Fig. 3.1: Photograph showing the Left Scapula of Ox- A. lateral view, B. Medial view

1. Supraspinous fossa,
2. Infra spinous fossa
3. Scapular spine
4. Acromion process
5. Cranial border
6. Caudal border
7. Cranial angle
8. Caudal angle
9. Glenoid angle
10. Neck
11. Subscapular fossa
12. Area serrata
13. Supra glenoid tubercle
14. Coracoid process
15. Glenoid cavity
16. Dorsal border

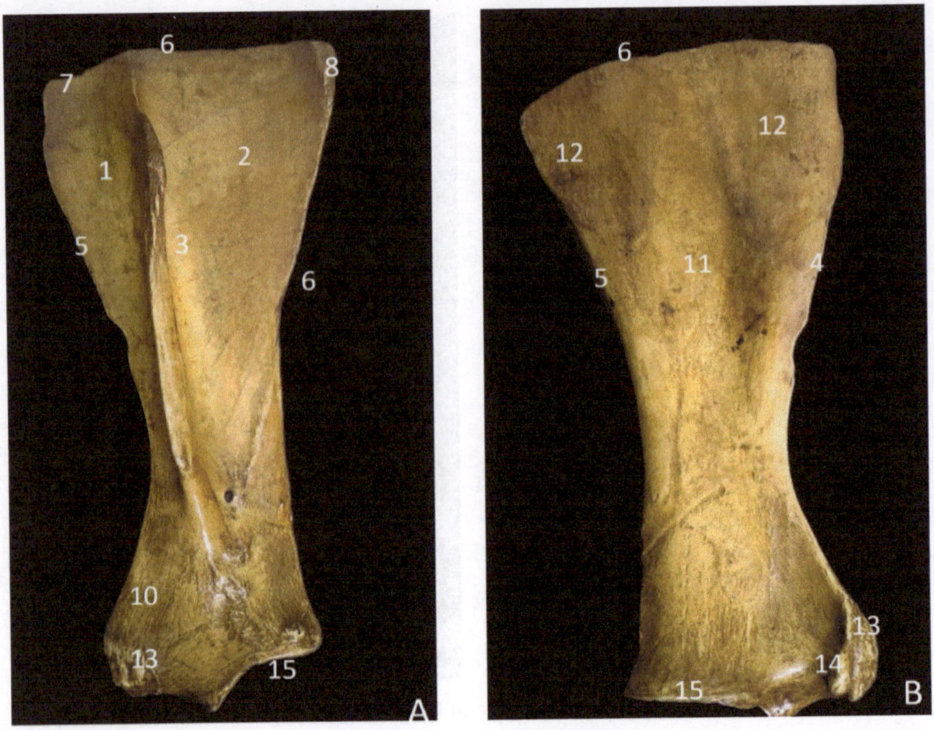

Fig. 3.2: Photograph showing the Left Scapula of Horse- A. lateral view, B. Medial view

1. Supraspinous fossa,
2. Infra spinous fossa
3. Scapular spine
4. Cranial border
5. Caudal border
6. Dorsal border
7. Cranial angle
8. Caudal angle
9. Glenoid angle
10. Neck
11. Subscapular fossa
12. Area serrata
13. Supra glenoid tubercle
14. Coracoid process
15. Glenoid cavity

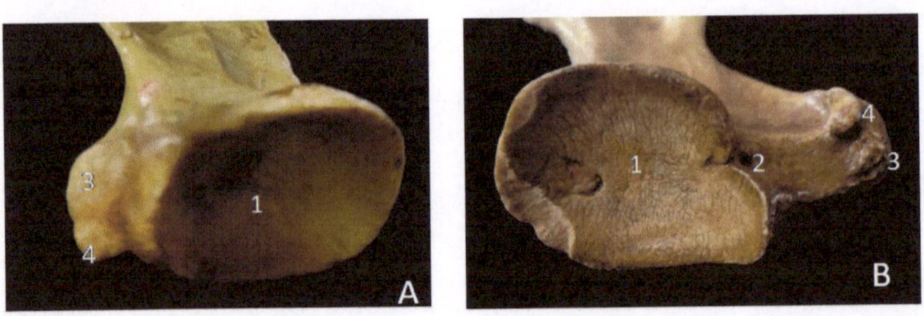

Fig. 3.3: Photograph showing Glenoid cavity of Ox (A) and Horse (B) 1. Glenoid cavity, 2. glenoid notch, 3. Supraglenoid tubercle, 4. Coracoid process

Table 3.1: Comparative Table on Scapula of Ox, Horse, Dog and Pig

SCAPULA				
Characters	**Ox**	**Horse**	**Dog**	**Pig**
Shape	More regularly triangular than horse with a broad dorsal area	Less regularly triangular than ox	Relatively long and narrow, Triangular in shape	Very wide and triangular
Ratio of the Supraspinous and Infraspinous fossae	1:3	1:2	1:1	1:0.7
Scapular cartilage	Extensive and semilunar shaped	Same as ox	In the form of thin band	Not extensive as in ox, shape like the blade of scythe
Scapular spine	More prominent	Roughly triangular in outline	Increase in height from before backward and reaches upto the rim of glenoid cavity	Triangular, wide at its middle, bears large tuberosity at the middle
Tuber spinae	Not distinct	Variably prominent, situated little dorsal to the middle of spine	Absent	Triangular situated little dorsal to the middle of spine
Acromion process	Present in the form of pointed projection	Absent	short, blunt and overhang the rim of glenoid cavity	rudimentary
Subscapular fossa	Shallower than horse	Deeper than ox	Shallow	Deeply excavated
Area serrata	Indistinct	Distinct and Deeper than ox	Shallower than horse	Shallow than horse
Glenoid cavity	Circular in shape	Oval in shape	Oval in shape	Oval in shape
Glenoid notch	present on antero-lateral aspect of the rim of glenoid cavity	deep and more distinct than ox, present on antero-medial aspect of rim of glenoid cavity	Absent	Absent
Supraglenoid tubercle	Smaller than horse	Larger than ox and placed further away from glenoid cavity	Blunt and located at the cranio-dorsal aspect of the glenoid cavity	Blunt, present on cranio-medial aspect of rim of glenoid cavity
Infraglenoid tubercle	Absent	Absent	Located distally at the thicker caudal border	Absent
Coracoid process	Short and rounded	Well developed	Absent	Not distinct/ absent

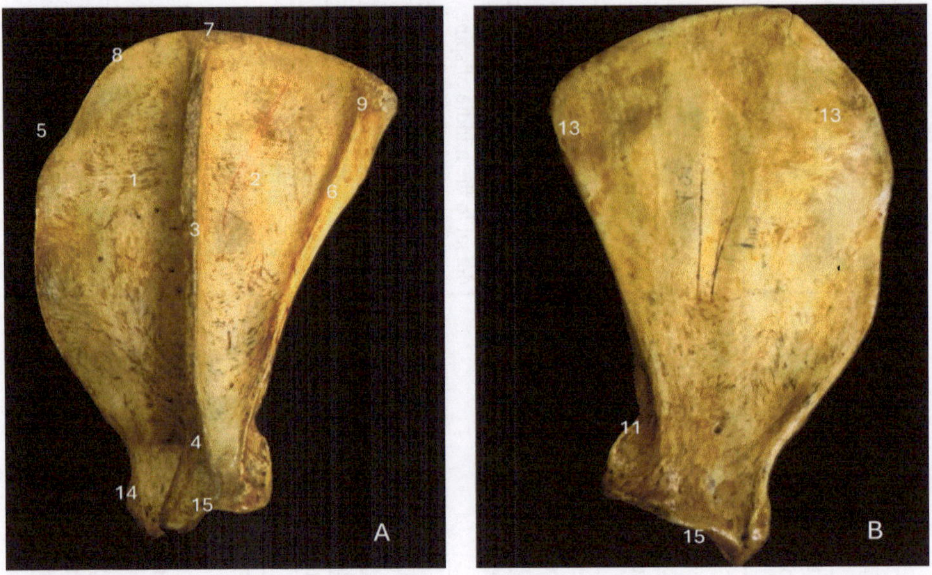

Fig. 3.4: Photograph showing the Left Scapula of Dog- A. lateral view, B. Medial view
1. Supraspinous fossa,2.Infra spinous fossa, 3. Scapular spine,4. Acromion process, 5. Cranial border, 6. Caudal border, 7. Dorsal border, 8. Cranial angle, 9. Caudal angle, 10. Glenoid angle, 11. Neck, 12. Subscapular fossa, 13. Area serrata, 14. Supra glenoid tubercle, 15. Glenoid cavity

Fig. 3.5: Photograph showing the Right Scapula of Pig lateral view
1. Supraspinous fossa, 2. infra spinous fossa, 3. Scapular spine, 4. Tuber spinae, 5. Cranial border, 6. Caudal border, 7. Dorsal border, 8. Cranial angle, 9. Caudal angle, 10. Glenoid angle 11. Neck, 12. Supra glenoid tubercle

b. Comparative Study of Humerus

Humerus of Ox

Humerus is the long bone of arm region which obliquely downward and backward (distally caudally) forming about 55° angle with horizontal plane. It forms shoulder joint above with the scapula and elbow joint below with radius and ulna. For description it presents a body and two extremities: proximal and distal (Fig. 3.6).

The Body is irregularly cylindrical and twisted in appearance and has four surfaces *viz.* Cranial, caudal, lateral and medial surface. The lateral surface is smooth and presents musculo-spiral groove which lodges *Brachialis muscle.* Medial surface is straight rounded from side to side and merged with cranial and caudal surfaces. Tubercle is located just above the middle of the surface which gives attachment to *Latissimus dorsi* and *Teres major muscles*. Cranial (Anterior) surface is triangular, wide and smooth above while narrow and rough below. It is separated from lateral surface by crest of humerus. Well distinct deltoid tuberosity is present on the middle of crest of humerus which gives attachment to *Deltoid muscle.* A rough line curves upward backward from deltoid tuberosity towards the neck to which *Lateral head of triceps muscle* is attached. Caudal surface is smooth and rounded from side to side and has nutrient foramen at distal third (Fig. 3.6).

Proximal extremity is large in size and comprised of head, neck, two tuberosity and bicipital / intertuberal groove. Head is circularly convex, twice larger than the glenoid cavity of scapula to which it articulate. A fossa is situated in front of head with many foramina. Neck is well defined caudally but practically absent elsewhere. Lateral tuberosity is large and prominent situated craniolaterally. It has two parts, a summit and a convexity. Summit is situated cranially and form lateral boundary of bicipital groove. It is separated from medial tuberosity by intertuberal/ bicipital groove. Medial tuberosity is smaller and consists of cranial and caudal parts. The cranial part forms the medial boundary of bicipital groove, whereas, caudal part is low in level and gives attachment to subscapularis muscle. Cranial part of the lateral and medial tuberosities gives attachment to *supraspinatus muscle*. Distal to summit a prominent oval rough area, *facies musculi infraspinati* is present which gives attachment to *infraspinatus muscle* (Fig. 3.6). Caudal part of lateral tuberosity gives attachment to *infraspinatus muscle*. Distal extremity has cranially two oblique, unequal surfaces for articulation with condyles of radius and separated by a ridge. Medial condyle (trochlea) is large in size, crossed by sagittal groove. It presents synovial fossa and caudally extended upto olecranon fossa for semilunar notch of ulna. Lateral condyle (capitulum) is small, oblique in appearance and has wide shallow groove. Radial (Coronoid) fossa is situated infront above the groove on the medial condyle. It receives the coronoid process of the radius during incomplete flexion of elbow joint. Condyloid crest forms the boundary of musculospiral groove. Olecranon fossa is situated behind for articulation with semilunar notch. Medial and lateral epicondyles are situated behind and above the condyles as two thick ridges projections on respective side. Lateral epicondyle had a rough raised area forming the cranial boundary of musculospiral groove. Medial epicondyle is more prominent and furnishes the origin of *flexor muscles of carpus and digit*. Whereas, the lateral epicondyles gives origin to *Extensor carpi radialis muscle.*

Table 3.2: Comparative table on humerus of Ox, Horse, dog And Pig

Characters	Ox	Horse	Dog	Pig
Shaft/ body	Irregularly cylindrical and twisted in appearance	Irregularly cylindrical and twisted in appearance	Long and slender in appearance	Triangular in proximal third and oval in distal third
Head	Circular ,convex	Circular, convex massive than ox	long and strongly curved from before backward	Strongly curved
Neck	Ill defined	Ill defined	More developed than ox	Well defined than ox
Medial tuberosity	Divided into cranial and caudal parts	Same as ox	Small, nodular and undivided	Single, round and elongated
Lateral tuberosity	Prominent, divided and raised about 3 cm above the level of head	Undivided, arched inward and doesn't raised as high as in ox	undivided and extended a little proximal to the level of head	Large, divided into two parts
Deltoid tuberosity	Less prominent	Prominent than ox	Low long ridge like	Smaller than ox
Bicipital groove	Undivided	Divided by intermediate tubercle/ ridge	Broad and undivided	Undivided and converted into canal
Facies musculi infraspinati	Prominent round rough area	Irregularly triangular area	In the form of shallow depression	Appears as an oval area
Musculospiral groove	Shallower than horse	Deeper than ox	Shallow as compare to ox and horse	Very shallow
Nutrient foramen	In the distal third of caudal surface	In the distal third of caudal surface	About middle of the caudal surface	Caudal surface below its middle
Teres tubercle	Prominent	Prominent	Present in the form of slight elevation	Absent
Olecranon fossa	Deeper and wider than horse	Shallower than ox	Deep and communicate with radial fossa	Narrow and deep
Radial fossa	Deeper and wider than horse	Shallower than ox	Shallow and communicate with olecranon fossa	Deep and prominent
Supra trochlear foramen	Absent	Absent	Present between radial and olecranon fossa	Occasionally present
Condyles	Medial condyle is larger than lateral condyle	Same as in ox	Lateral condyle is convex and articulate with radius , medial condyle is deeply concave and articulate with ulna	Condyles are equal in size

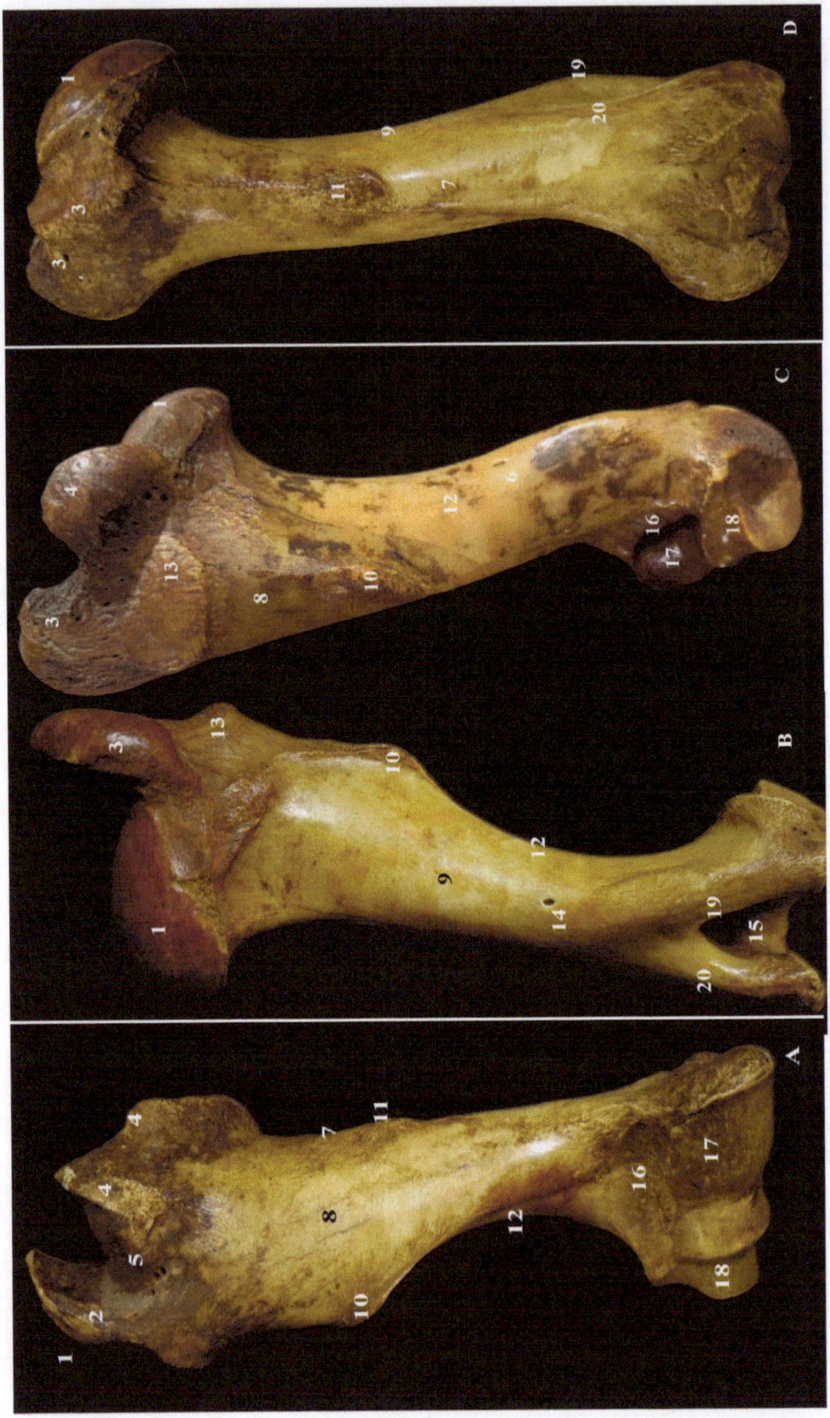

Fig. 3.6: Photograph showing the Right Humerus of Ox: A. Cranial view, B. Caudal view, C. Lateral View and D. Medial view
1. Head, 2. Lateral tuberosity (summit), 3. Lateral tuberosity (convexity), 4. Medial tuberosity, 5. Bicipital groove, 6. Lateral surface, 7. Medial surface, 8. Cranial surface, 9. Caudal surface, 10. Deltoid tuberosity, 11. Teres tubercle, 12. Musculospiral groove, 13. Facies musculi infraspinati, 14. Nutrient foramen, 15. Olecranon fossa, 16. Radial fossa, 17. Lateral condyle, 18. Medial condyle, 19. Lateral Epicondyle, 20. Medial epicondyle

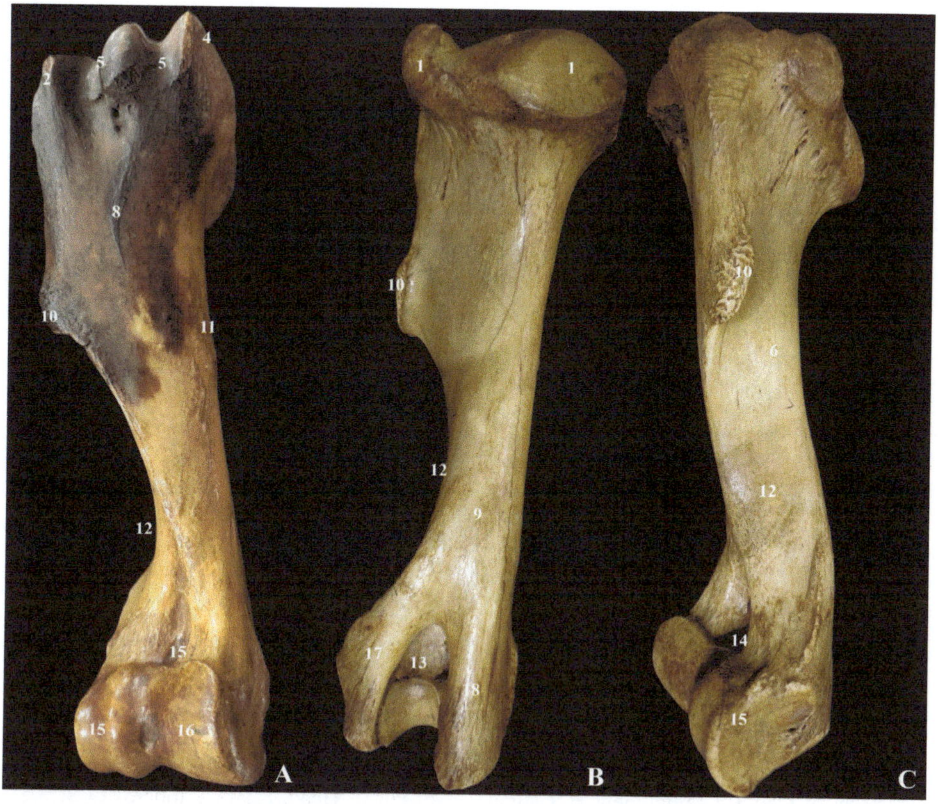

Fig. 3.7: Photograph showing the left Humerus of Horse. A. Cranial view, B. Caudal view and C. Lateral View

1. Head
2. Lateral tuberosity (summit)
3. Lateral tuberosity (convexity)
4. Medial tuberosity
5. Bicipital groove
6. Lateral surface
7. Medial surface
8. Cranial surface
9. Caudal surface
10. Deltoid tuberosity
11. Teres tubercle
12. Musculospiral groove
13. Olecranon fossa
14. Radial fossa
15. Lateral condyle
16. Medial condyle
17. Lateral epicondyle
18. Medial epicondyle

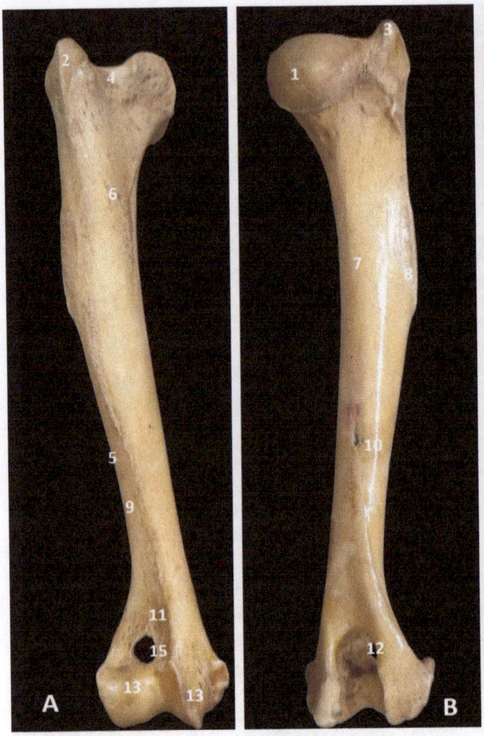

Fig. 3.8: Photograph showing the Right Humerus of Dog A. Cranial view, B. Caudal view

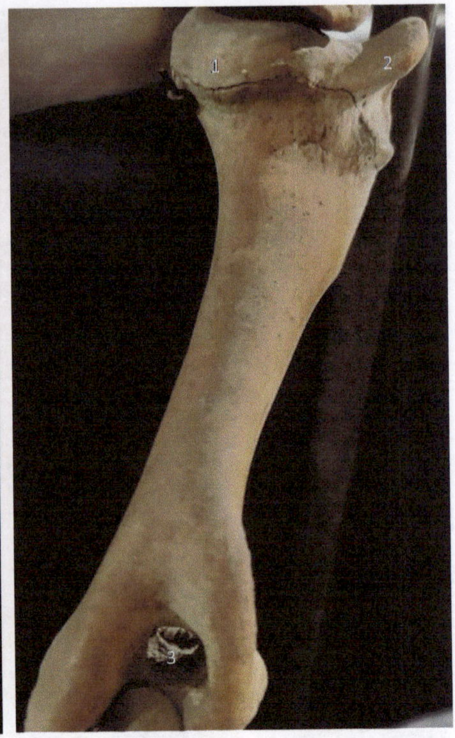

Fig. 3.9: Photograph showing theHumerus of Pig: 1. Head, 2. lateral tuberosity, 3. supra trochlear foramen

1. Head
2. Lateral tuberosity
3. Medial tuberosity
4. Bicipital groove
5. Lateral surface
6. Cranial surface
7. Caudal surface
8. Deltoid tuberosity
9. Musculospiral groove
10. Nutrient foramen
11. Olecranon fossa
12. Radial fossa
13. Lateral condyle
14. Medial condyle
15. Supra trochlear foramen

c. Comparative Study of Radius and Ulna

Radius and Ulna of Ox

Forearm region is formed by the Radius and ulna.

Radius: It is long and slightly curved bone directed vertically and forms elbow joint with humerus above and carpal joint with carpal bones below. It presents a shaft/ body and two extremities proximal and distal extremities.

The **Shaft** is flattened from before backward in the proximal part and curved longitudinally. Proximally it consists of two surfaces (cranial and caudal) but in distal part it has lateral smooth surface which is narrow above and wide below. So it become prismatic lower down where it bears cranial, caudal and lateral surfaces (Fig. 3.10).

Cranial surface is convex, smooth and rounded side to side and covered with *extensors of carpus and digit*. **Caudal** surface is concave attached with ulna except at two interosseous spaces. A groove connects these spaces laterally. **Lateral border** is rounded in its proximal part and slightly curved, wide and flat below. It gives attachment to *Lateral digital extensor and Extensor carpi obliques muscles*. **Medial border** is concave and mainly subcutaneous. Long medial ligaments are attached on this border. Proximally a rough area is present on this border for *Brachialis muscle*. In the middle it presents rough area separated from posterior surface by a ridge (Fig. 3.10).

The **Proximal extremity** is flattened from before backward and wide transversely. It presents the humeral articular surfaces (glenoid cavities) divided by a sagittal groove which corresponds to that of distal end of humerus. The medial articular area is larger than lateral one. It presents a synovial fossa which extends medially form the deep groove between the two glenoid cavities. It is crossed by sagittal ridge which ends infront into a prominent lip, the coronoid process. Caudally, it presents two facets for articulation with ulna. A slightly elevated rough area, Radial/ bicipital tuberosity is situated medially for insertion of tendon of *Biceps brachii muscle*. Medial tuberosity is continuous with radial tuberosity. Lateral tuberosity is well marked and gives attachment to *Lateral digital extensors*. Medial and lateral tuberosities are for attachment of collateral ligaments (Fig. 3.10).

The **Distal extremity** is large and thick medially. It articulate with proximal row of carpal bones and presents three oblique facets (Fig. 3.10). Medial facet is largest, quadrangular and concavoconvex for articulation with radial carpal bone. Intermediate facet is similar in form but smaller than medial one and articulate with intermediate carpal bone. Lateral facet is smaller that above two facets and partially furnishes by ulna for articulation with ulnar carpal bone (Fig. 3.10). Cranial surface of the distal extremity presents three grooves separated by ridges. The middle and lateral ridges are large and similar which gives passage to tendon of *Extensar carpi radialis* muscle and *Common digital extensor muscles*, respectively. The medial ridge is small, oblique and gives passage to *Extensor carpi obliquus* muscle.

Table 3.3: Comparative Table on Radius of Ox, Horse, Dog and Pig

RADIUS				
Characters	**Ox**	**Horse**	**Dog**	**Pig**
Thickness	Radius is thicker than ulna	Radius is thicker than ulna	Radius is thicker than ulna	Ulna is thicker than radius
Proximal interosseous space	Present	Present	Present	Present
Distal interosseous space	Present	Absent as ulna reached upto the middle of the radius	Present	Present
Radial tuberosity	Present in the form of rough elevation	Well marked than ox	Present in the form of Small rough protuberance	Rough area
Facet for ulna	Large facet is present	Present and smaller than ox	Present	Present
Styloid process	Absent	Absent	Present	present
Articular facets on distal extremity	Oblique in direction	Straight than ox	Wider forming trochlea an extensive concave carpal articular area	Concavo- convex
carpal bones articulation	Radial, intermediate and ulnar carpal bones	Radial, intermediate, ulnar and accessory carpal bones	Ulnar and intermediate carpal bones	Radial and intermediate carpal bones

Ulna : It is ill developed long bone fused with radius along its caudolateral aspect. It presents shaft, and proximal and distal extremities (Fig. 3.10). The proximal part of the body/ shaft contains a medullary cavity which extends somewhat into the proximal end. Shaft/body is roughly prismatic and presents three surfaces and three borders. Cranial surface is fused with caudolateral aspect of radius except at proximal and distal interosseus spaces. Medial surface is smooth and convex. Lateral surface is convex. Medial and lateral borders are thin and sharp except in the interosseus space. The caudal border is concave and smooth, sharp below and projected towards the distal end. Major part of the bone is formed by the proximal extremity, which projects upward and backward slightly behind the distal end of humerus to form point of elbow. It contains two surfaces and two borders. The cranial border bears on its middle a pointed notch, the semilunar notch which is triangular in outline, concave from above downward and convex transversely, and articulates with the humerus (Fig. 3.10). A synovial fossa is present in its lower part. The caudal border is smooth and concave. Free end or summit is tuberous, referred as olecranon to which *Triceps muscle* attached. Cranial border near termination presents distinct beak like projection, Anconean process. Distal extremity is fused with radius and form lateral facet for ulnar carpal bone which projects below the radius known as styloid process of ulna (Fig. 3.10).

Table 3.4: Comparative Table on Ulna of Ox, Horse, Dog and Pig

Characters	Ox	Horse	Dog	Pig
Body	Three sided, curved	Three sided	Comparatively Large, three sided, more rounded dorsally, cross caudal surface of radius mediolaterally	Curved, placed more medially behind the radial tuberosity
Olecranon	Large, bears rounded tuberosity	Very Massive	Olecranon is grooved presents three tubercles, caudal one is large and rounded	Same as in ox undivided
Anconeus process	In the form of Pointed projection	less prominent than ox	highly projected hook like.	Same as in ox
Semilunar notch	triangular , concave from above downward, convex, transversely	Same as in ox	Same as ox but wide distally	Same as ox
Styloid process	Present	Absent	Present and articulates with ulnar and accessory carpals	Present

Characters	Ox	Horse	Dog	Pig
Articulation with radius	Ossified	Ossified	Articulate proximally and distally. A complete inetosseous space present between radius and ulna	Firmly united with radius by dense fibrous tissue
Articulation with carpal	Ulnar carpal	No articulation with carpal	Ulnar carpal	Ulnar carpal

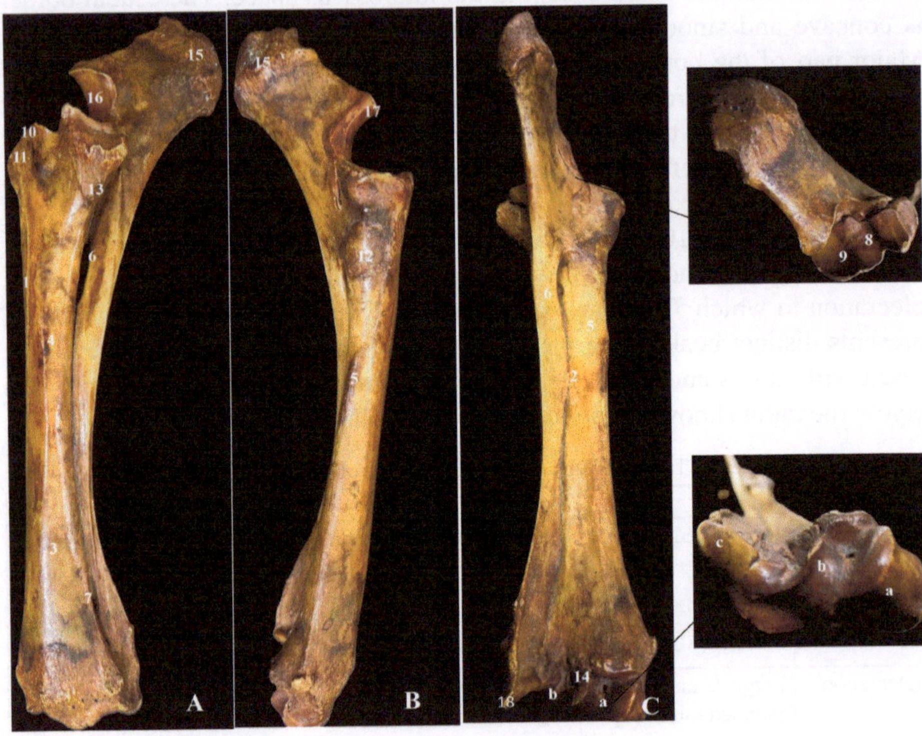

Fig. 3.10: Photograph showing the Left Radius and Ulna of Ox-A. Cranio-lateral view, B. Medial view and C. Caudal view

Radius of ox
1. Cranial surface
2. Caudal surface
3. Lateral surface
4. Lateral border
5. Medial border
6. Proximal interosseus space
7. Distal interosseus space
8. Sagittal groove
9. Synovial fossa
10. Coronoid process
11. Radial tuberosity
12. Medial tuberosity
13. Lateral tuberosity Articular facets for
    a. Radial carpal
    b. Intermediate carpal
    c. Ulnar carpal Ulna of Ox
15. Olecrenon tuberosity
16. Semilunar notch
17. Anconeus process
18. Styloid process

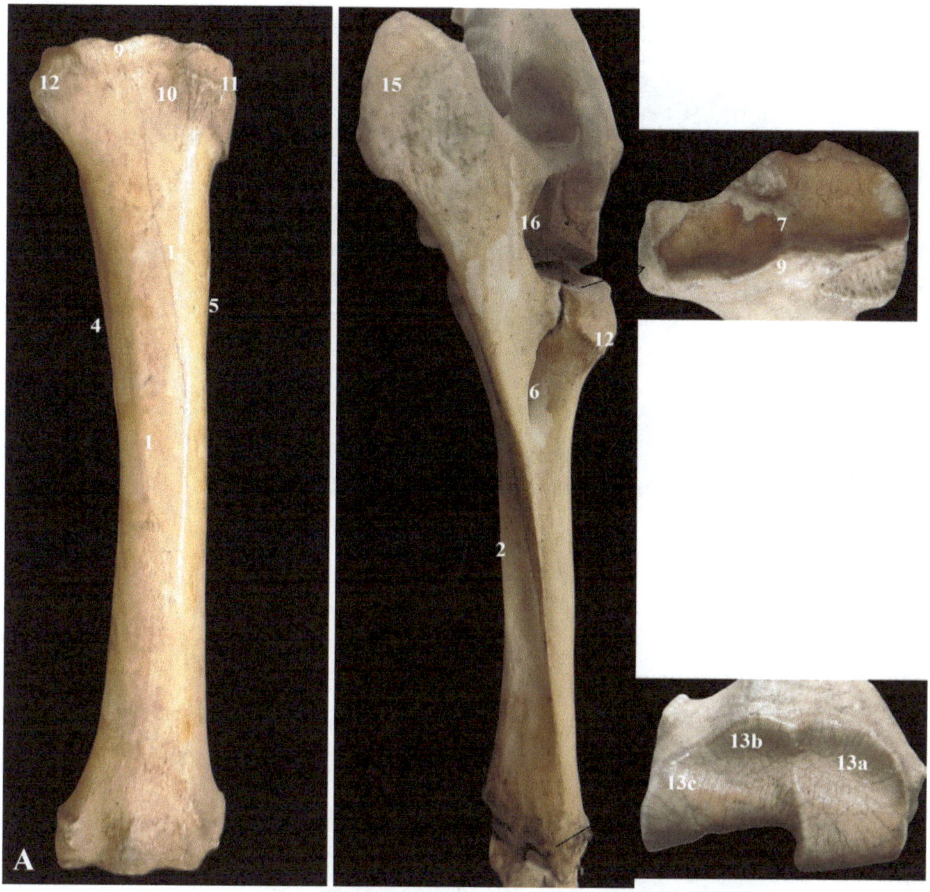

Fig. 3.11: Photograph showing the Right Horse A. Cranial View of Radius, B. Caudo-lateral view of Radius and Ulna

1. Cranial surface
2. Caudal surface
3. Lateral surface
4. Lateral border
5. Medial border
6. Proximal interosseus space
7. Sagittal groove
8. Synovial fossa
9. Coronoid process
10. Radial tuberosity
11. Medial tuberosity
12. Lateral tuberosity
13. Articular facets
    a. For radial carpal
    b. Intermediate carpal
    c. Ulnar carpal

Ulna of Horse

15. Olecrenon tuberosity
16. Semilunar notch

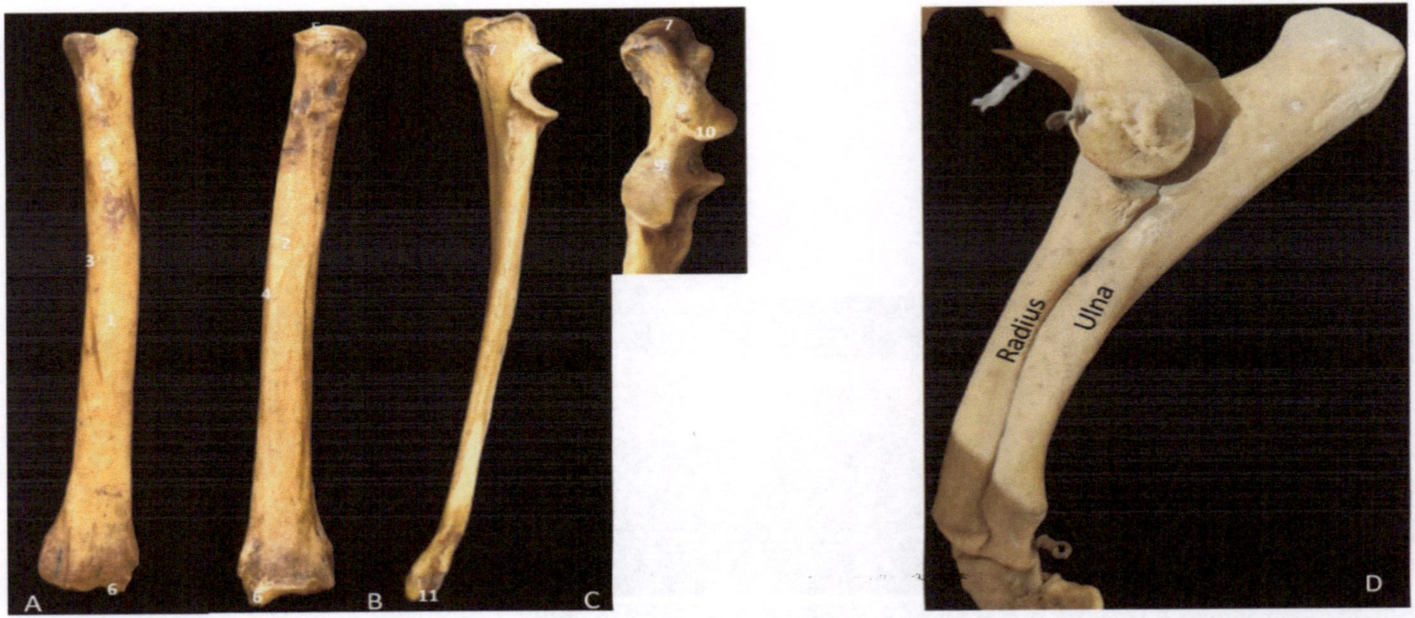

Fig. 3.12: Photograph showing the Radius and Ulna of Dog (A,B and C) and Pig (D) A. Cranial view of radius, B. Caudal View of radius, C. Ulna lateral view

1. Cranial surface
2. Caudal surface
3. Lateral border
4 Medial border
5 Facet for ulna
6 Styloid process
7. Olecrenon tuberosity
8. Olecranon proess
9. Semilunar notch
10. Anconeus process
11. Styloid process

d. Comparative Study of Carpals

Manus Region of the Body

The manus region of the body is formed by the carpals, large and small metacarpal bones, third and fourth digits and proximal and distal sesamoid bones.

Carpal bones: Carpals are designates the bones of knee joint. Carpals are eight short bones arranged in two rows between radioulna above and metacarpal below. The proximal row consist of four carpal bones *viz.* Radial/ Os scaphoideum, intermediate/ Os lunatum, ulnar/ Os triquetrum and accessory/ Os pisiforme carpals, while distal row are designated numerically as first/ Os Trapezium, second/ Os trapezoideum, third/ Os capitulum and fourth carpals/ Os hamatum.

Table 3.5: Showing the Arrangement of Carpals in Different Species

Species	Rows	Arrangement of carpal bones			
Ox	Proximal row	Radial	Intermediate	Ulnar	Accessory
	Distal row	Second and third fused carpal		Fourth carpal	
Horse	Proximal row	Radial	Intermediate	Ulnar	Accessory
	Distal row	First	Second	Third	Fourth
Dog	Proximal row	Radial and Intermediate fused carpal		Ulnar	Accessory
	Distal row	First	Second	Third	Fourth
Pig	Proximal row	Radial	Intermediate	Ulnar	Accessory
	Distal row	First carpal	Second	Third	Fourth

Carpals of Ox

In ox the proximal row has 4 carpals *viz.* radial, intermediate, ulnar and accessory carpals, whereas, distal row presents two carpal bones. In distal row first carpal is absent, second and third is fused to form 2nd and 3rd fused carpal and fourth carpals are (Fig. 3.13).

Proximal row

Radial / carpal Scaphoid / Os carpi radiale/ Os scaphoideum: It is the largest bone of the proximal row (Fig. 3.13). It is situated on medial side of the proximal row and presents six surfaces. Proximal surface is convexo concave and articulate with distal extremity of radius. Distal surface is convex infront and concave behind. It articulates with second and third fused carpal bone. Dorsal, palmar and medial surfaces are rough and non articular. Palmar surface is tuberculated. Lateral surface has three facets, one larger above and two smaller below. An excavated area present between them. These facets articulates with medial surface of intermedial carpal bone (Fig. 3.13).

Intermediate carpal / carpal Semilunar / Os carpi intermedium/ Os lunatum: It is wedge shaped bone, wider infront and constricted in the middle. Its proximal surface is convexo-concave and articulates with radius (Fig. 3.10). The distal surface articulates with second and third fused and fourth carpal bones. Medial surface have three reciprocal facets for articulation with lateral surface of radial carpal bone. Lateral surface has two facets which articulate with corresponding facets of ulnar carpal bones. Dorsal and palmar surfaces are rough and nonarticular. Palmar surface has upward facing and inward facing two prominence (Fig. 3.13).

Ulnar carpal / carpal Cuneiform / Os carpi ulnare/ Os triquetrum: It is very irregular bone with long axis directed obliquely backward and laterally. It has six surfaces. Proximal surface is extensive and sinuous and articulates with both radius and ulna. Distal surface is oblique and undulating for articulation with the 4th carpal bone. Medial surface has two facets - an upper the other lower, for articulation with the corresponding facets on the lateral surface of the intermediate carpal bone. Dorsal and lateral Surface are convex, rough and non articular. Palmar surface is oblique. It presents large oval facets for articulation with the accessory carpal.

Accessory carpal / Pisiform / Os carpi accessorium/ Os pisiforme: It is short, thick and roughly rounded piece of bone. Its dorsal surface has concave facet for ulnar carpal bone (Fig. 3.10).

Carpals of distal row: in distal row first carpal bone is absent or in constant.

Second and third fused carpal / Osmagnum/Os trapezoideum: tThe proximal surface has two unequal facets separated by an dorso-Palmar ridge. The medial facets is larger and concavo-convex and articulates with the distal surface of the radial carpal while lateral one is the small and articulates with the distal surface of the intermediate carpal. The Palmar surface is continued by the proximal surface. The lateral surface has three facets for articulation with the corresponding facets on the medial surfaces of the 4th carpal. Distal surface is extensive and almost plain with a synovial fossa in the centre. It articulates with the large metacarpal bone (Fig. 3.13).

Fourth carpal / Unciform: It is quadrilateral piece of bone (Fig. 3.10). The proximal surface has two facets which are separated by a median ridge. The facets articulate with the distal surfaces of the intermediate and ulnar carpal bones. The distal surface has large facet for articulates with the large metacarpal (4th component) bone. The medial surface has three facets for articulation with the corresponding facets on the lateral surface on the combined second and third carpal bone. The dorsal and palmar surfaces are rough and non-articular.

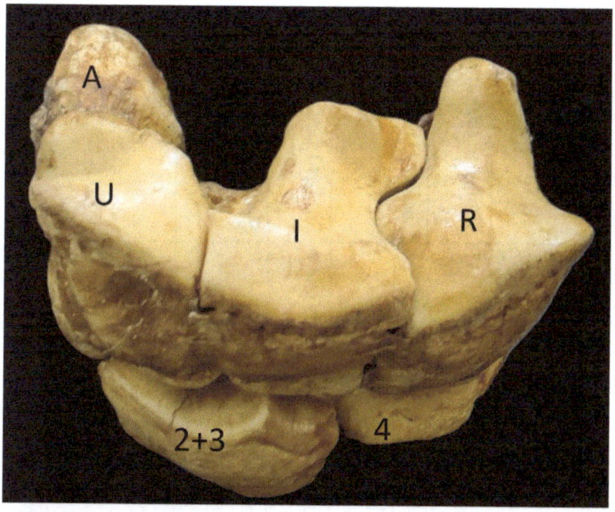

Bones of Proximal row

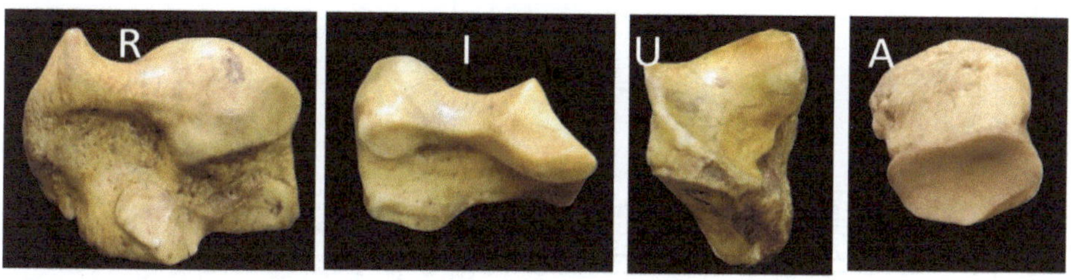

Bones of distal row

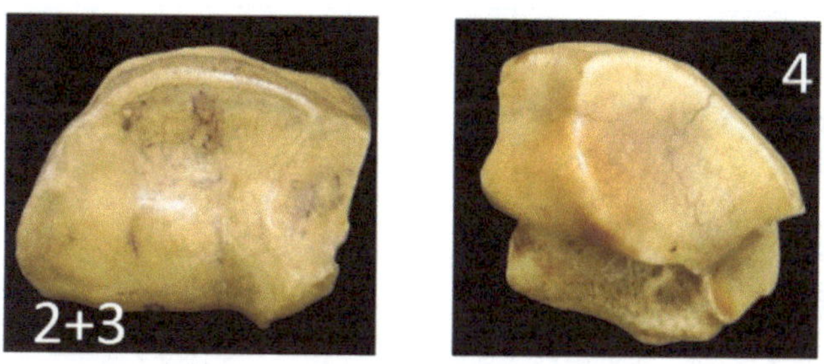

Fig. 3.13: Photograph showing the Carpals of OX (Proximal view)

Carpals of Ox
R- radial carpal
I- Intermediate carpal
A- Accessory carpal
U- Ulnar Carpal
2+3- second and third fused carpal
4- Fourth carpal

Table 3.6: Comparative Table on Carpal of Ox, Horse, Dog and Pig

Characters	Ox	Horse	Dog	Pig
Radial carpal 1. Shape	Less regular than horse Long axis is obliquely caudo-medially directed	resembles with ox but it is more regular, long axis is not oblique	The radial and intermediate carpals are fused to form a large bone which articulates with almost the entire distal surface of the radius and with the bones of the distal row.	Similar to ox
2. Proximal surface	Narrower than horse and posteriorly curved upward	Wider than ox, Does not curve upward behind.		Similar to ox
Medial surface	Rough and non- articular	Same as in ox		Same as in ox
Lateral surface	Has three facets for articulation with intermediate carpal	Same as in ox		Same as in ox
Intermediate carpal 1. shape	Regular in shape, Constricted at its middle and wider behind than front	More regular than ox, wedge in shape, wider in front and proximal surface is saddle shaped.		Similar to ox
2. Long axis	directed obliquely caudo-medially	long axis is not oblique		Similar to ox
3. External angle	Less pointed	More pointed.		Similar to ox
Ulnar carpal 1. shape-	Large and very irregular	It is the smallest bone in the proximal row.	It is long	Similar to ox
2. proximal surface	Articulates with both radius and ulna	Concave, which articulates with the radius only	Articulates with the radius and ulna above	Similar to ox
3. distal articulation	4th carpal	4th carpal	4th carpal and prolonged downward to articulate with fifth metacarpal also.	Similar to ox
4.palmar surface	Not oblique, bears large oval facet	Oblique and bears concave facet and a tubercle distallt		

Characters	Ox	Horse	Dog	Pig
Accessory carpal 1. shape	Short, thick and rounded.	Discoid and has two surfaces and a circumference. Lateral surface is convex and rough. The medial surface is concave and forms the lateral wall of the carpal groove.	Cylindrical, constricted in its middle and enlarged at each end.	Resembles with horse but has no lateral groove.
2. Articulation	Articulates with ulnar carpal only	The dorsal border bears two facets for articulation with radius (upper one) and ulnar carpal bone.	It articulates with the ulna and the ulnar carpal bone.	articulate with ulnar carpal bone
First carpal	Absent	It is inconstant, When present it is situated behind the second carpal bone. Shape and size is about a pea	It is the smallest bone of the distal row. It articulates with the second carpal laterally and first metacarpal distally.	Small, elongated from before backward, rounded articulate infront with second carpal bone.
Second carpal 1. Shape	Fused with 3rd carpal to form large quadrilateral bone. The proximal surface articulates with the distal surface of the radial and intermediate carpal Distal surface is extensive and articulates with the large metacarpal bone.	Smallest bone of the distal row. irregularly hemispherical in shape.	It is a wedge shaped bone. Its proximal surface is Its	High and narrow
2. Proximal surface		Articulates with the radial carpal	convex	
3. lateral surface		Bears three facets for articulation with the third carpal bone.		
4. distal surface		Articulates with the second metacarpal bone and third or large metacarpal bone.	Concave, rests on the second metacarpal bone.	articulates with 2nd and 3rd metacarpal
Third carpal		This is the largest bone among the carpal bone of distal row.	It is somewhat similar to the second carpal bone.	

Characters	Ox	Horse	Dog	Pig
1. Proximal surface		Articulate with the distal surface of the radial carpal and the intermediate carpal bones.		Articulates with radial and intermediate carpal
2.distal surface		Articulate with the large metacarpal and the second metacarpal bone.	It distal surface is concave which articulates with the third metacarpal bone.	articulates with 3rd metacarpal
3. medial surface		Three facets for articulation with the second carpal bone.		
4. lateral surface		Two facets for articulation with medial surface of the 4th the carpal bone.		
Fourth carpal 1. Shape	Smaller quadrilateral bone.	It is wedge shaped having palmar tubercle	The largest bone of the distal row.	The largest bone of the distal row.
2. Proximal surface	Articulate with the distal surfaces of the intermediate and ulnar carpal bones.	Convex and articulates with the intermediate and the ulnar carpal bones.		Articulates with intermediate and ulnar carpal.
3.Distal surface	Large facet for articulates with the large metacarpal (4th component) bone.	Articulates with the large and outer small metacarpal bones.	Articulates with the fourth and fifth metacarpal bone.	articulate with 4tha and 5tha metacarpal
4. Medial surface	Three facets for articulation with lateral surface on the combined second and third carpal bone.	Three facets for articulation with the third carpal bone.		

Name of carpal	Ox	Horse
Radial carpal	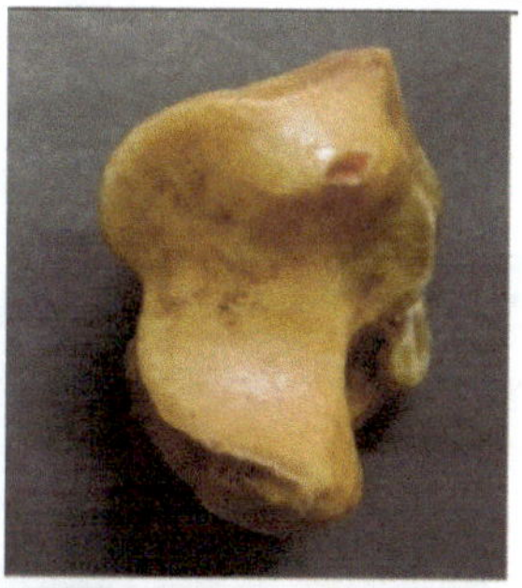	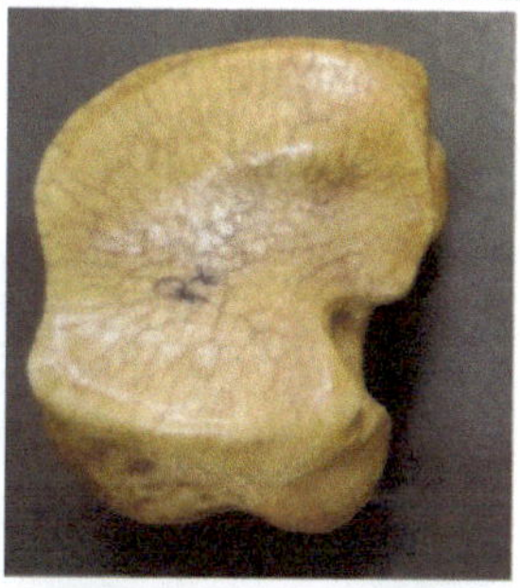
Intermediate carpal	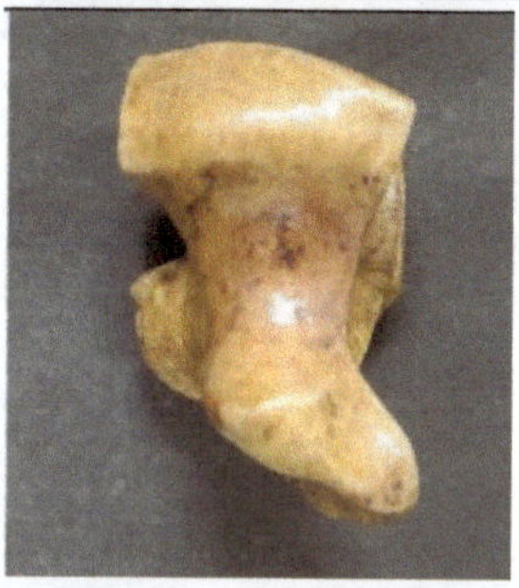	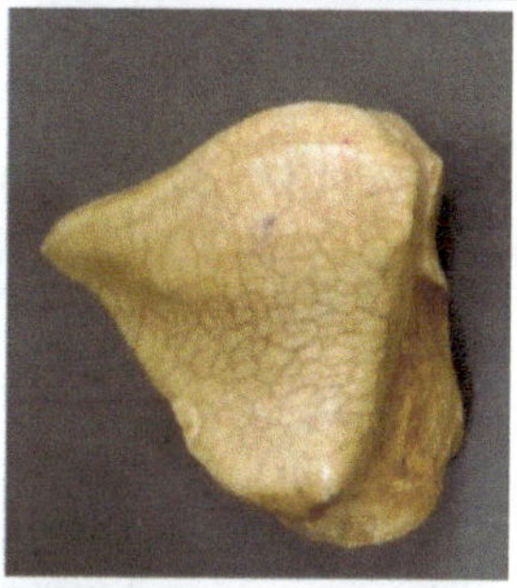
Ulnar carpal	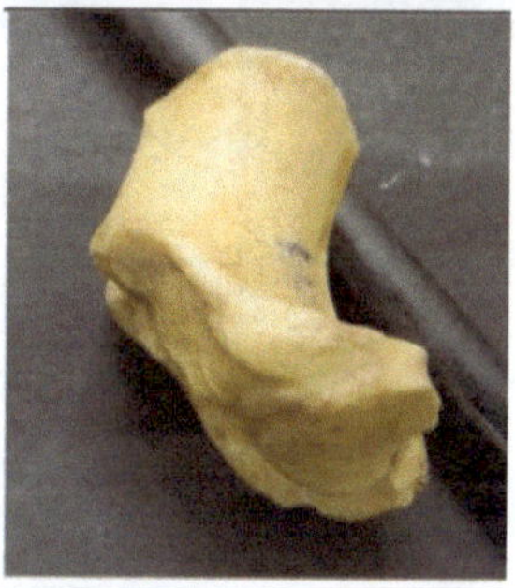	
Accessory carpal		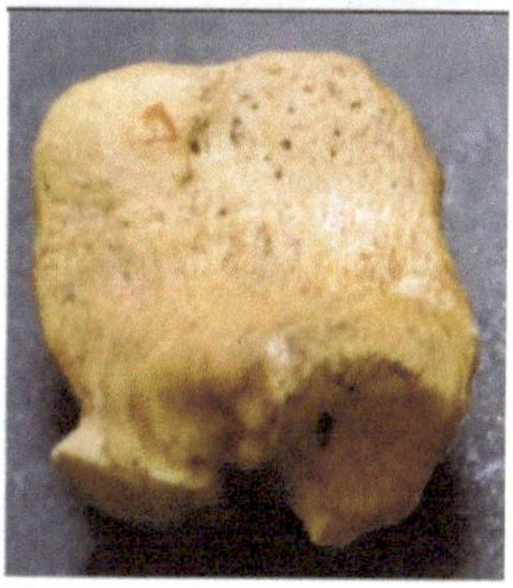

<table>
<tr><td>2nd carpal</td><td rowspan="2">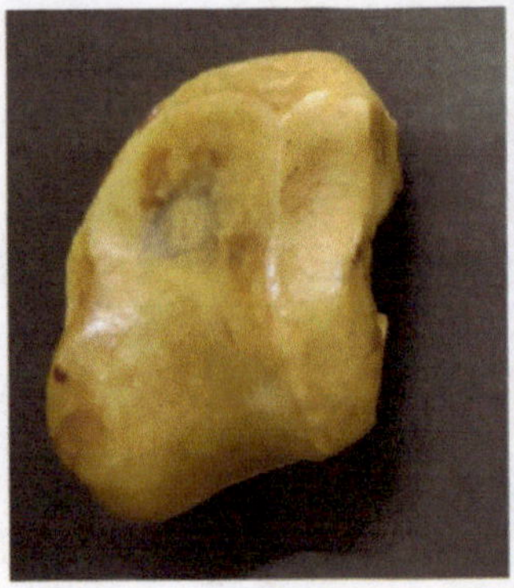</td><td></td></tr>
<tr><td>3rd carpal</td><td>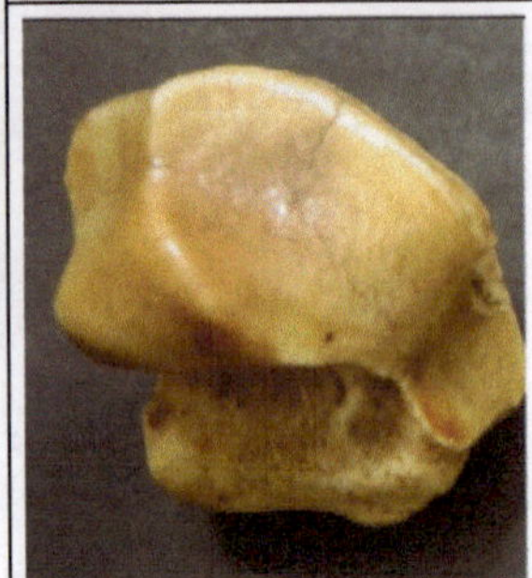</td></tr>
<tr><td>4th carpal</td><td>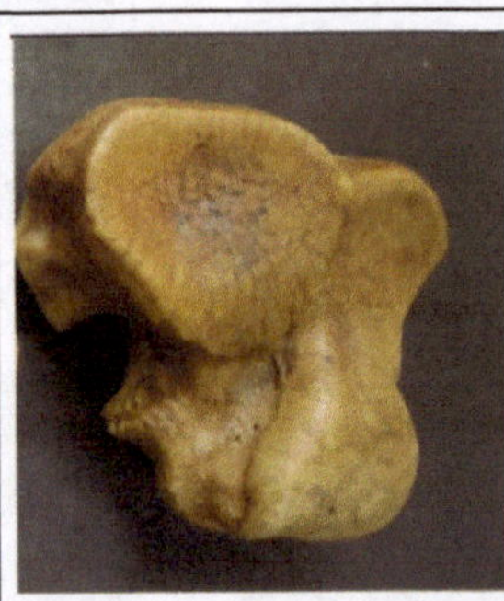</td><td>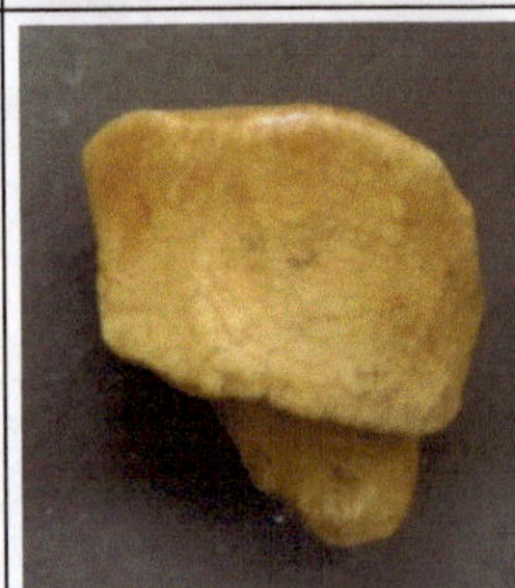</td></tr>
</table>

Fig. 3.14: Comparative photographs showing the difference between carpals of Ox and Horse (proximal view)

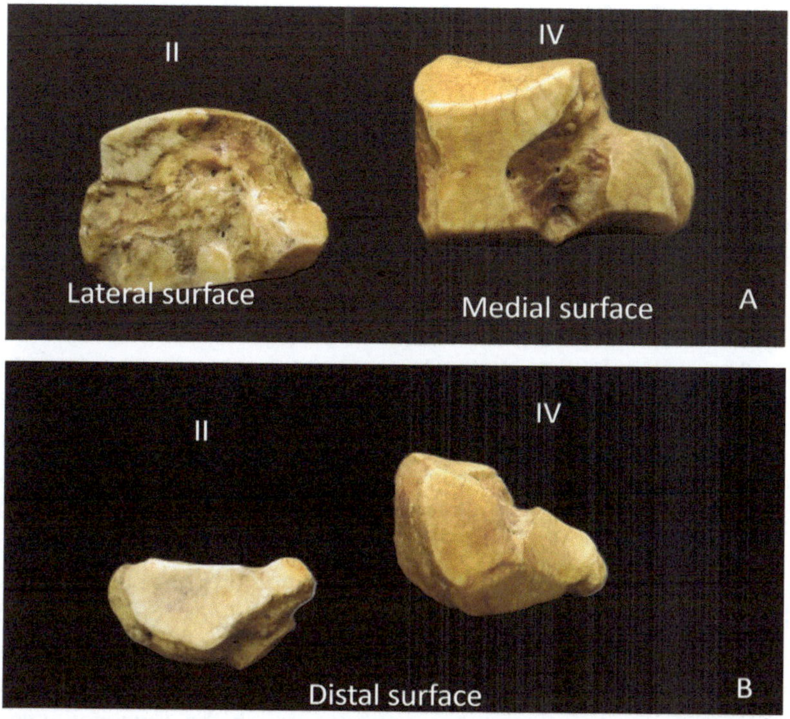

Fig. 3.15: Photograph showing the Difference between II and IV carpal of Horse

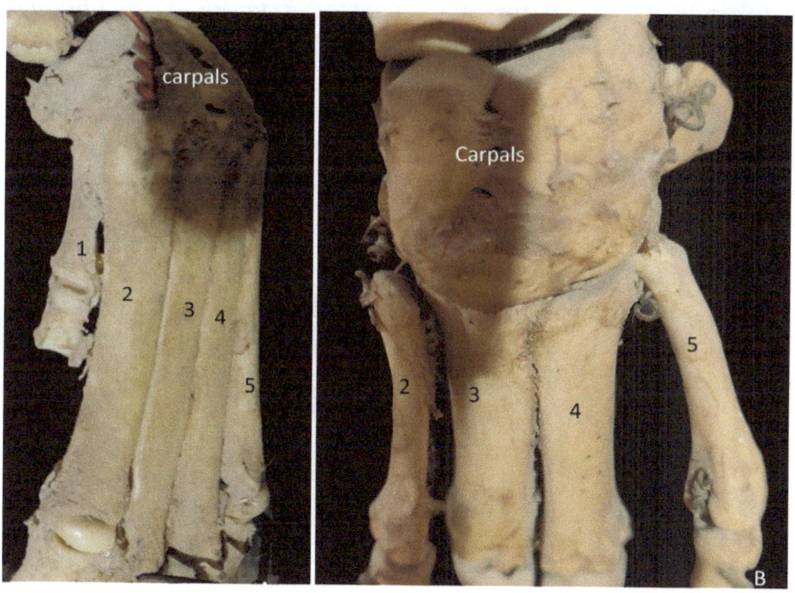

Fig. 3.16: Photograph showing A. Metacarpal of Dog, B. Carpal and Metacarpal of Pig

e. Comparative Study of Metacarpal

Metacarpal of Ox

Large metacarpal: - It is very strong long bone which is situated vertically between the carpus above and the proximal phalanx below. In ox, 3rd and 4th metacarpals are fused to form large metacarpal bone known as cannon which carries well developed 3rd and 4th functional digits. The union of 3rd and 4th metacarpal bones are marked by groove and has double medullary cavity. It presents a shaft; and proximal and distal extremities. The shaft/ body is irregularly cylindrical, flattened and presents two surfaces and two borders. The dorsal surface is smooth, rounded and convex from side to side and nearly straight in its length. It is marked by a vertical vascular groove or dorsal metacarpal groove (FIg. 3.17). Palmar surface is flat and also presents a similar but much fainter groove. This surface lodges the suspensory ligament of fetlock joint. Two rounded borders; lateral and medial are rough in the proximal third and smooth lower down. The proximal extremity has two slightly concave facets which are separated by a ridge in front and a notch behind (FIg. 3.17). The medial facet is larger than lateral one which articulates with the distal surface of the fused second and third carpal bones. The lateral facet is smaller and articulates with the fourth carpal bone. The medial part of the extremity has cranial tuberosity, meta-carpal tuberosity into which the *Extensor carpi radialis* muscle is inserted. Similarly, at the caudo- medial aspect of proximal end a small tubercle is present for attachment of *Extensor carpi obliques* muscle. The lateral aspect of caudal surface bears a small facet for articulation with small metacarpal bone (5th metacarpal). Distal extremity is divided into two parts by a sagittal notch. Each part is made up of two condyles which are separated by a sagittal ridge. The medial condyle is slightly larger than the lateral one. Each part articulates with the proximal surface of proximal phalanx below, and two proximal sesamoid bones behind.

Small metacarpal: Fifth metacarpal is the aborted long bone known as small metacarpal of ox. It is situated on the caudolateral side of large metacarpal. It is in the form of rounded rod and about one and half inch in length. Its proximal end articulates with large metacarpal by a small facet present on anterior surface. The distal end is pointed (FIg. 3.17).

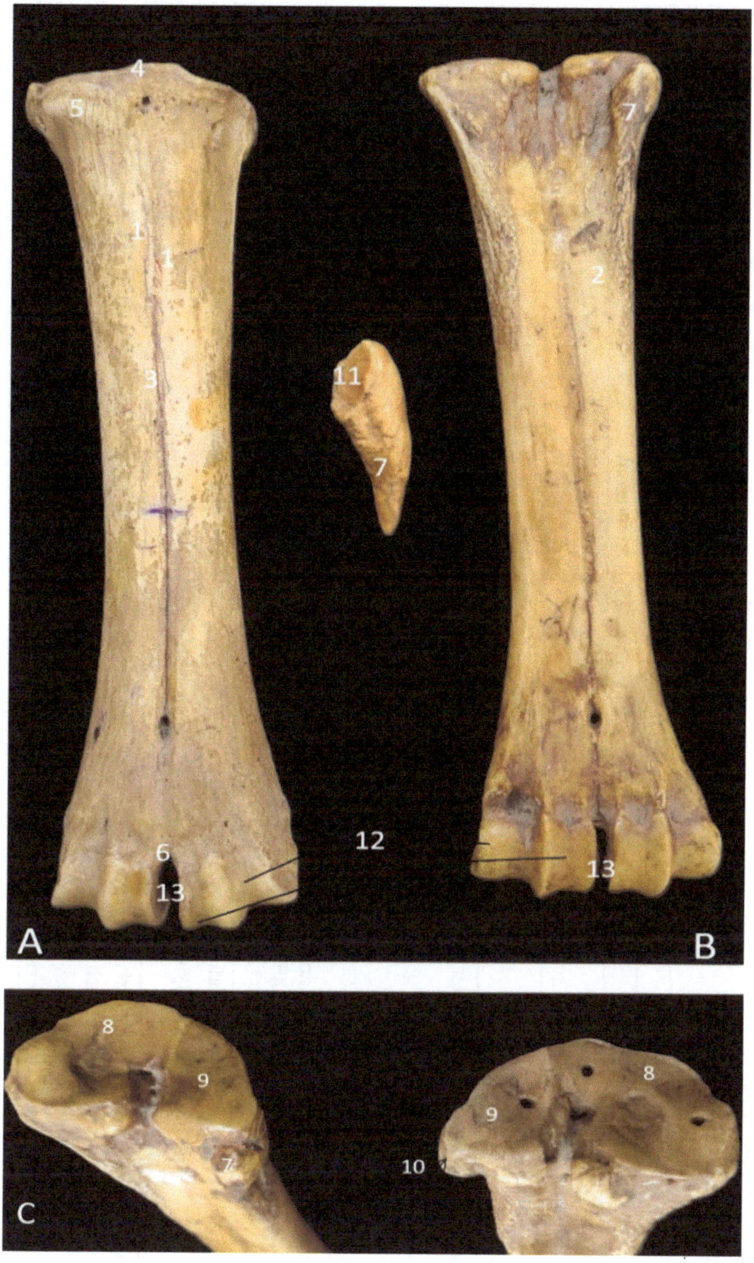

Fig. 3.17: Photograph showing the Metacarpal of Ox: A. Dorsal view, B. Palmar view, C. Proximal extremity, 1. **Dorsal surface,** 2. **Palmar surface,** 3. **Dorsal metacarpal groove,** 4. **Proximal extremity,** 5. **Metacarpal tuberosity,** 6. **Distal extremity,** 7. **Small metacarpal,** 8. Facet for 2nd & 3rd fused carpal, 9. Facet for 4th carpal, 10. Facet for small metacarpal, 11. Facet for large metacarpal, 12. Condyle, 13. Sagittal cleft

Table 3.7: Comparative Table on Metacarpal of Ox, Horse, Dog and Pig

Characters	Ox	Horse	Dog	Pig
Total no of metacarpal	3+4fused large metacarpal and 5TH small metacarpal	3rd large metacarpal and 2nd and 4th are small metacarpal	1st ,2nd, 3rd, 4th and 5th are separate metacarpal	2nd, 3rd, 4th and 5th are separate metacarpal
Functional (carries a digit)	3 and 4	3	1,2,3,4,5	2,3,4,5
Non-functional	5th	2,4 (Splint bone)	None	None
Absent	1st and 2nd	1st and 5th	All are present	1st
No. of large metacarpal	One (3+4 fused)	One (3rd)	All five are large	3 and 4
No of small metacarpal	One (5th)	Two (2nd and 4th)	Absent	2 and 5 (much smaller)
Shape of the body	Wider, flat and shorter than horse	Semi cylindrical longer than ox	Compressed dorso palmarly 1st- shortest and slender 3&4- four sided, rectangular 2 &5 –three sided	3 & 4- Three sided, longer and three times thicker than small metacarpal
Articular facet at the proximal extremity	Two facets for articulation with 2nd and 3rd fused carpal and 4th carpal bones	Three facets for articulation with 2nd ,3rd and 4th carpal	Corresponding carpal bone	Articulates with each other and 2,3,4 carpal
Metacarpal tuberosity	Present cranio-medially at proximal extremity of large metacarpal	Same as in ox	Present in 2nd and 3rd metacarpal	Present in 3rd metacarpal
Distal extremity	Divided into two parts by sagittal notch, each contain two condyles separated by sagittal ridge	Appears as one side of the ox, two condyles separated by ridge - Articular area are larger than ox	Head like articular surface with sagittal ridge palmarly	It appears as condyloid dorsally and trochlear palmarly

Characters	Ox	Horse	Dog	Pig
Small metacarpal	5th, located caudo-lateral	2 (caudo-medial) and 4 (caudo-lateral)	Absent	2 and 5 are smaller than 3rd and 4th
a. Shape and Size	Rounded rod 3.5-4 cm	Three sided, reached 2/3 to ¾ length of large metacarpal	3rd and 4th are triangular whereas 2nd and 5th are four sided	
b .Articulation	Only with large metacarpal	Large metacarpal and 2nd- two facets for articulation with 2&3 carpal, occasionally with 1st carpal, 4th - one facet for articulation with 4th carpal	2ndmetacarpal- 1st, 2nd and 3rd carpal an 1st and 3rd metacarpal 3rd metacarpal- 3rd and 4th carpal 4th and 5th metacarpal- 4th carpal	2nd and 5th metacarpal articulates with 3rd and 4th large metacarpals and 2nd and 4th carpals, respectively
c. Distal end	Pointed	Small nodular, reaches upto 2/3 or ¾ of large metacarpal	In the form of head	Condyloid

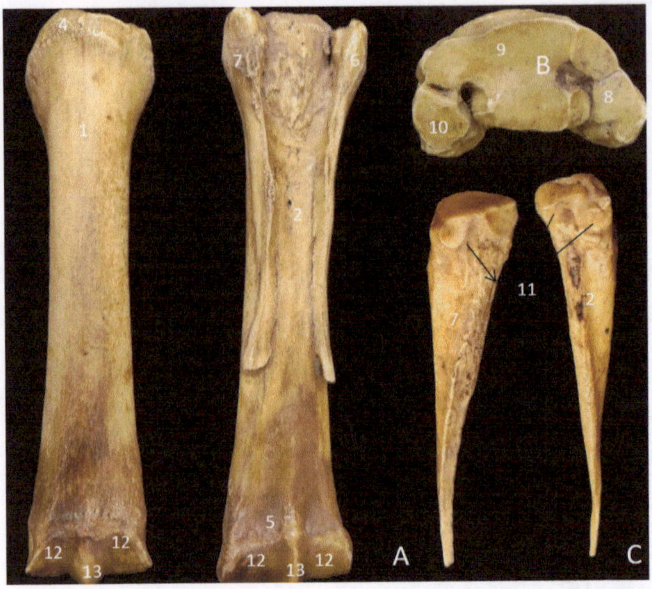

Fig. 3.18: Metacarpal of Horse A. 3rd Large metacarpal, B. Proximal extremity of large metacarpal C. 2nd and 4th Small metacarpals

1. Dorsal surface
2. Palmar surface
3. Proximal extremity
4. Metacarpal tuberosity
5. Distal extremity
6. 2nd Small metacarpal
7. 4th Small metacarpal
8. Facet for 2nd carpal
9. Facet for 3rd carpal
10. Facet for 4th carpal
11. Facet for large metacarpal
12. Condyle
13. Sagittal ridge

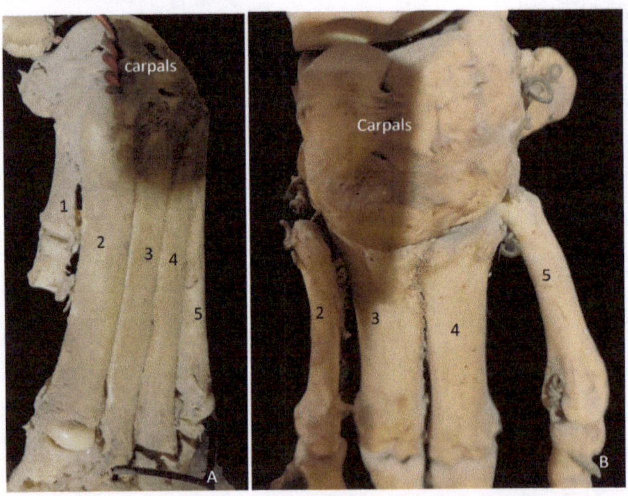

Fig. 3.19: Photograph showing A. Metacarpal of Dog, B. Carpal and Metacarpal of Pig

f. Comparative Study of Digits, Phalanges and Sesamoids

Digits of Ox

The ox has two fully developed (3rd and 4th) digits. First digit is absent and 2nd and 5th digits are rudimentary, non functional and form dew claws, which are situated caudally at metacarpophalangeal joint without articulation. Each developed digit has three phalanges (Fig. 3.20).

First phalanx/ os suffraginis: the bone appears small but it is a long bone having a shaft/ body, two extremities and a marrow cavity. The bone is directed obliquely downward and forward between metacarpal above and 2nd phalanx below. The body has four surfaces. Dorsal and Abaxial surfaces are continuous. Palmar surface presents two tubercles for attachment of sesamoidian ligaments. Axial/ interdigital surface is flat. Proximal end is broad and bear a concave articular surface divided into larger and higher abaxial and lower axial area by a groove. Posterior to these articular cavities two facets are present for articulation with sesamoid bone. Distal end is condyloid (abaxial condyle is larger) and articulate with proximal end of 2nd phalanx and form the pastern joint (Fig. 3.23).

Second phalanx/ os coronae: It is also a long bone, placed obliquely downward and forward directed and situated between first and third phalanx (Fig. 3.20). The distal half of 2nd phalanx is included within the hoof. The shaft is three sided. Axial surface is flat. Abaxial surface is larger and rough. Proximal end presents two concave glenoid cavities for the condyles of first phalanx. The abaxial glenoid cavity is larger than axial one. Palmar surface is encroached by distal articular surface. It distally has two tubercles for the attachment of the tendon of *Superficial digital flexar muscle*. The distal extremity is smaller and is divided by a sagittal groove into two condyles, the abaxial condyle is larger than axial one. Distal end articulates with the proximal surface of 3rd phalanx below and the distal sesamoid bone behind (Fig. 3.20 and 3.23).

Third phalanx/ Os pedis/ Coffin bones: It is a short bone, resembles the shape of hoof in which the whole bone is situated. It has four surfaces and an angle. The Abaxial surface is convex, rough slopes downward and forward and transverse by cranial-caudal groove (preplanter groove) at distal end with several perforations for vessels. Axial/ interdigital surface is rough above and smooth below and presents groove at distal end. At its proximal angle it is perforated by large palmar foramina which lead to a cavity in the middle of the bone. Palmar surface is slightly concave and flat, wide at middle and narrowest in front. It is also called solar surface. The proximal surface is narrow from side to side and slopes downward and backward and is oblique transversely

(Fig. 3.20 and 3.24). The interdigital side is lower. It presents two concavities for condyles which articulate with distal end of 2nd phalanx and form the coffin joint. A small facet is present on the posterior part of proximal surface for the sesamoid bone. The proximal or coronary border bears an eminence, the extensor process; to which tendon of *Common digital extensor muscle* is attached. The lateral border is sharp infront and rounded behind. It separates the abaxial surface and palmar surface. Sharp medial border separates the palmar surface from axial surface. The dorsal border is rounded and present between axial and abaxial surface. The angle is very short and blunt, which projects backward.

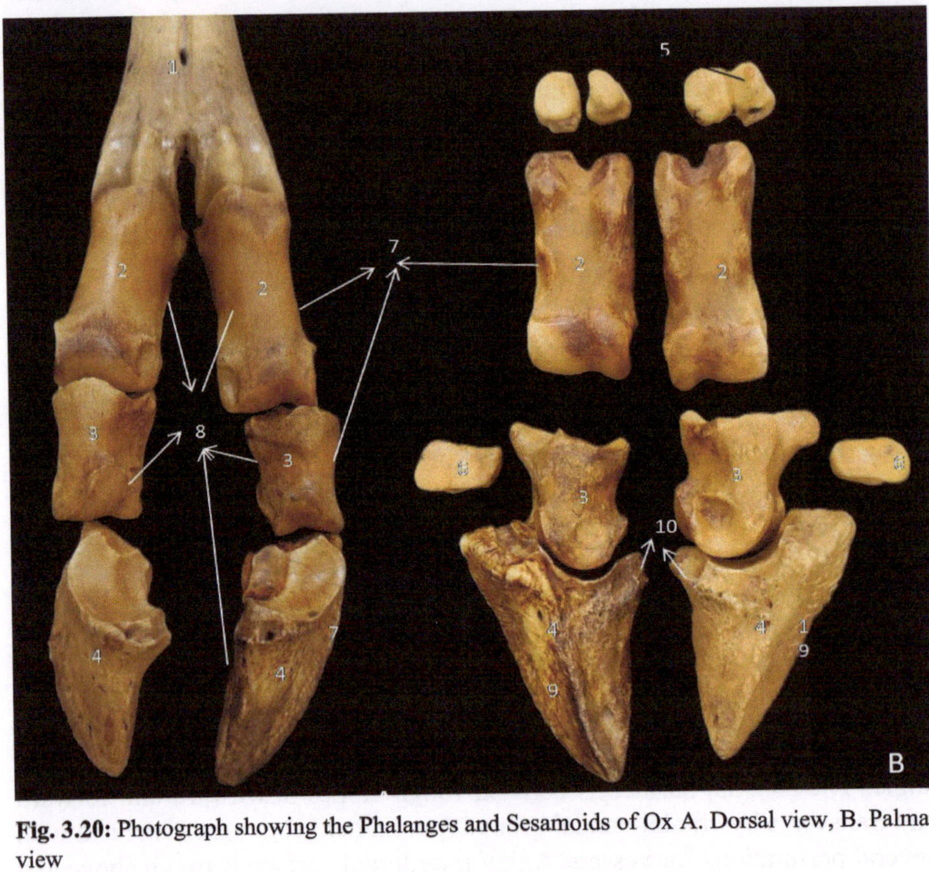

Fig. 3.20: Photograph showing the Phalanges and Sesamoids of Ox A. Dorsal view, B. Palmar view

1. Metacarpal
2. 1st phalanx
3. 2nd phalanx
4. 3rd phalanx
5. Proximal sesamoids
6. Distal sesamoids
7. Abaxial surface
8. Axial surface
9. Solar surface
10. Extensor process

Sesamoid bones

Proximal sesamoids: They are four in number in each limb in which axial sesamoid bone is larger than abaxial one. All are oval elongated. Each bone is three sided having a base and an apex. The apex is proximal and rounded. Dorsal surface is concave while palmar surface is convex. The base has a facet which articulate with corresponding facets at palmar aspect of distal end of first pair of phalanges (two in each digit) and the dorsal surface articulates with the corresponding part of distal end of the large metacarpal bone (Fig. 3.21).

Distal sesamoids: They are two in number, one for each digit. Each bone is flat and situated transversely between second and third phalanx. The middle part of the bone is wider than the end narrow part. Dorsal surface is articular faces upwards. It has a ridge at the middle and cancave articular surface flanked from this. It articulates with the palmar surface of 2nd phalanx. Palmar surface is flat and bear a ridge in the middle. Distal border bears infront a narrow facet for articulation with the distal or third phalanx (Fig. 3.21).

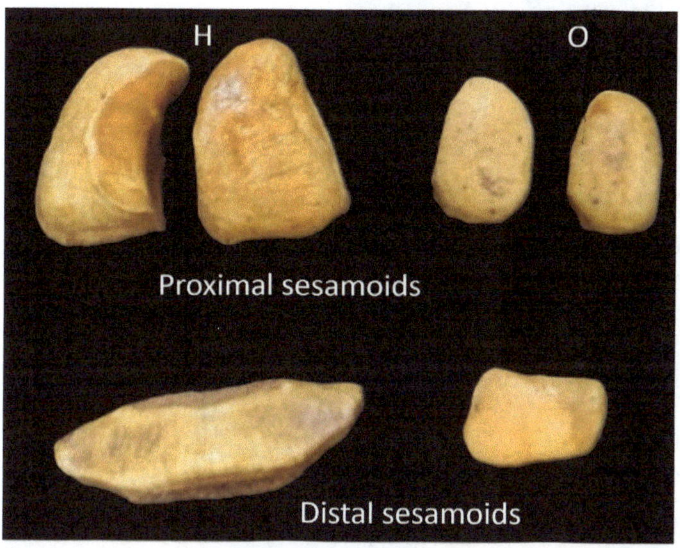

Fig. 3.21: Comparative photograph of Sesamoids of Horse (H) and ox (O)

Horse: There is only one digit i.e. manus consist of single I, II & III phalanx and three sesamoids. First phalanx is large and wide (Fig. 3.21). Dorsal surface is convex and smooth. Palmar surface is flat and marked by triangular rough area for distal sesamoidean ligament (Fig. 3.23). Second phalanx is short, dorso-palmarly compressed. Width is more than length. The shape of third phalanx is just like a hoof and is covered by it. It presents dorsal, solar and articular surfaces. Palmarly two prismatic masses, palmar processes (angles) are projected each side in which medial one is shorter. Each is divided into proximal and distal part by a notch which may be converted into a foramen or canal by bar of bone. There are two proximal and one distal (navicular) sesamoid bones. Proximal sesamoids are three sided pyramid shaped and distal sesamoid is shuttle shaped. These are larger than ox (Fig. 3.20 and 3.24).

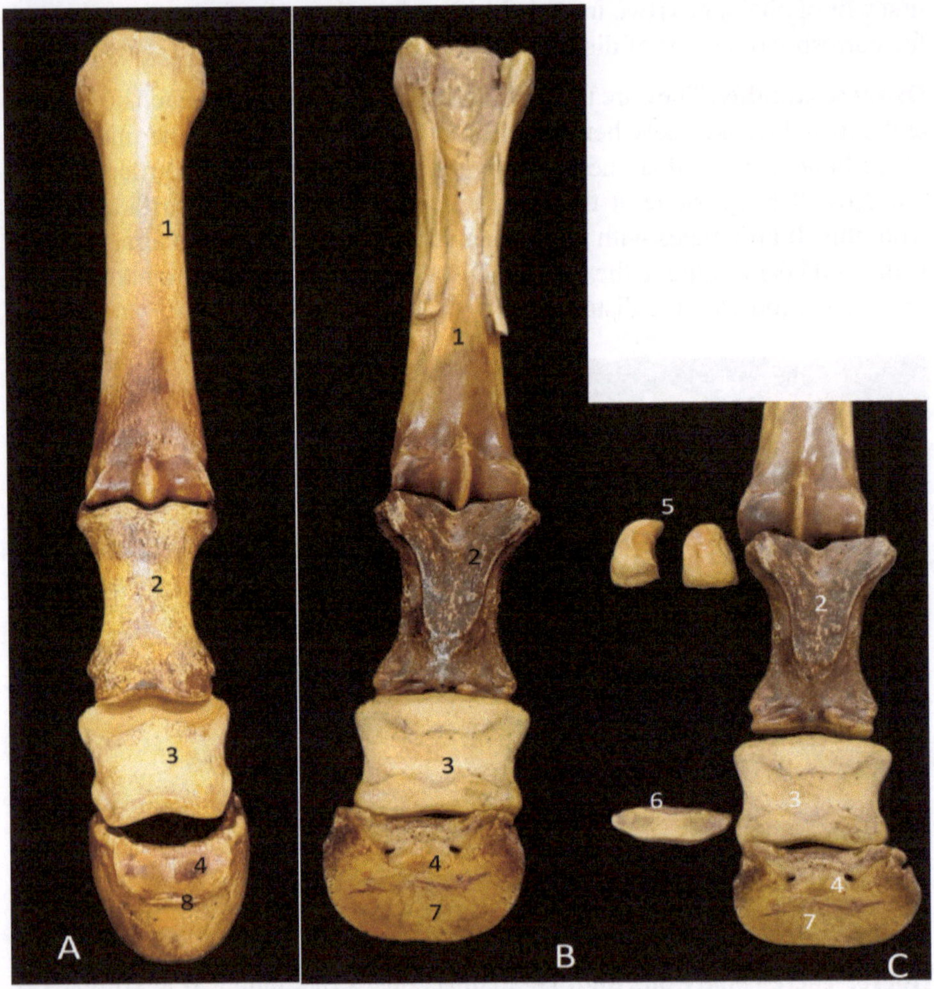

Fig. 3.22: Photograph showing the Metacarpal, phalanges and sesamoids of Horse A. Dorsal view, B. Palmar view, C. Along with Sesamoids

1. Metacarpal, 2. 1st phalanx, 3. 2nd phalanx, 4. 3rd phalanx, 5. Proximal sesamoids, 6. Distal sesamoids, 7. Solar surface, 8. Extensor process

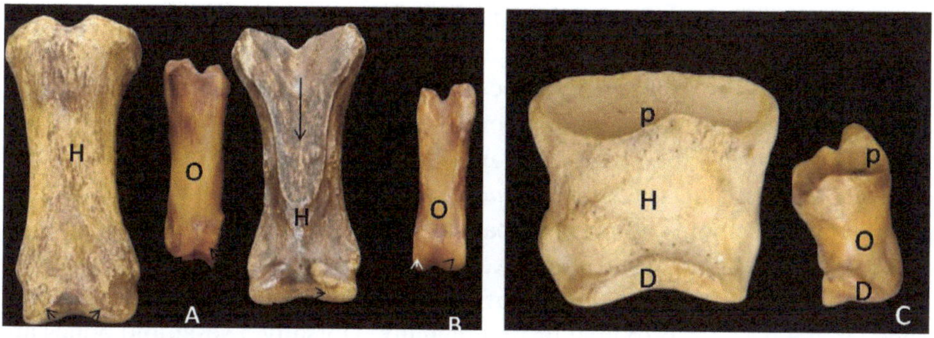

Fig. 3.23: Photograph showing the Comparison between 1st Phalanx (A and B) and 2nd Phalanx (C) of Ox(O) and Horse (H)

A. Dorsal surface, B. Palmar surface, P -Proximal surface, D-Distal surface arrow showing V shaped area in palmar surface in horse, arrow head- distal condyles

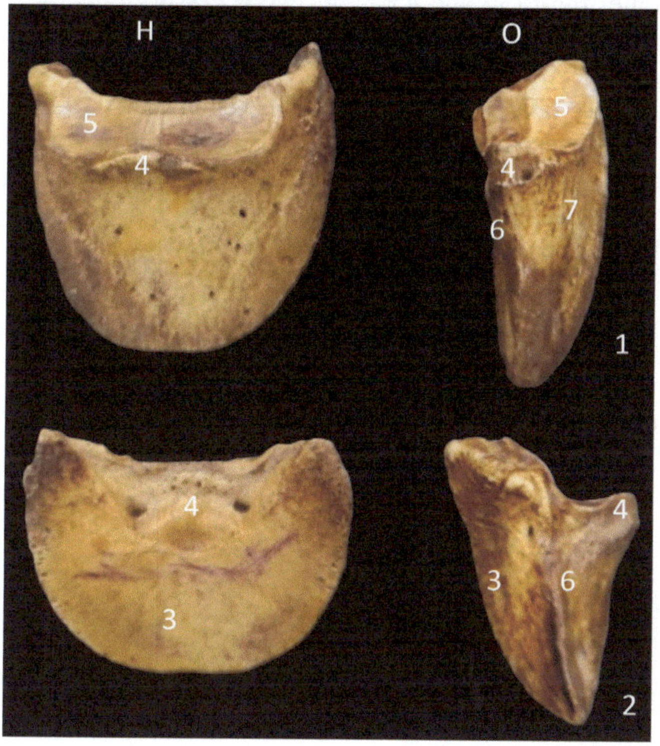

Fig. 3.24: Comparative photograph of 3rd Phalanges of Ox (O) and Horse (H)

1. 3rd phalanx (dorsal view)
2. 3rd phalanx (Palmar view)
3. Solar surface
4. Extensor process
5. Proximal surface
6. Axial surface
7. Abaxial surface

Dog: First digit is underdeveloped and possesses two phalanges. Other 4 digits have three phalanges. Third and fourth are larger. Each first phalanx has 4 surfaces. Proximal end is concave and distal end is trochleated. Second phalanx is short. Proximal end is concave and distal end is condyloid. Third phalanx resembles the shape of claw. Proximal end is articular and bear a bony border around it, *Crista unguicularis*. Distal end is pointed. There are 9 proximal sesamoids situated on palmar aspect (2 are placed posteriorly on each metacarpo-phalangeal joint except the first where only one is placed). They articulate with the distal end of metacarpal bone infront, and have small facet on the base for proximal or first phalanx. Dorsally 5 sesamoid bones are present (one occurs in the capsule of each metacarpo-phalangeal joints). Distal sesamoids are five in number cartilaginous. They are nodular in form and their size is slightly bigger than the pin head.

Pig: The pig has two chief digit (3rd and 4th) and two accessory digits (2nd and 5th). Each chief digit has 3 phalanges and 3 sesamoids. Chief digits are thicker and longer than the accessory digits. Proximal/ 1st phalanx of each digit is twice the length of 2nd phalanx. Extensor process is absent. Proximal sesamoids are present in between the metacarpo- phalangeal joint (two for each joint). Each sesamoid is similar to ox but are narrow and ridged behind. Phalanges of accessory digits are similar but smaller and do not reach the ground. Distal sesamoids are flat and oval. One distal sesamoid is present to each distal inter-phalangeal joint (Fig. 3.25).

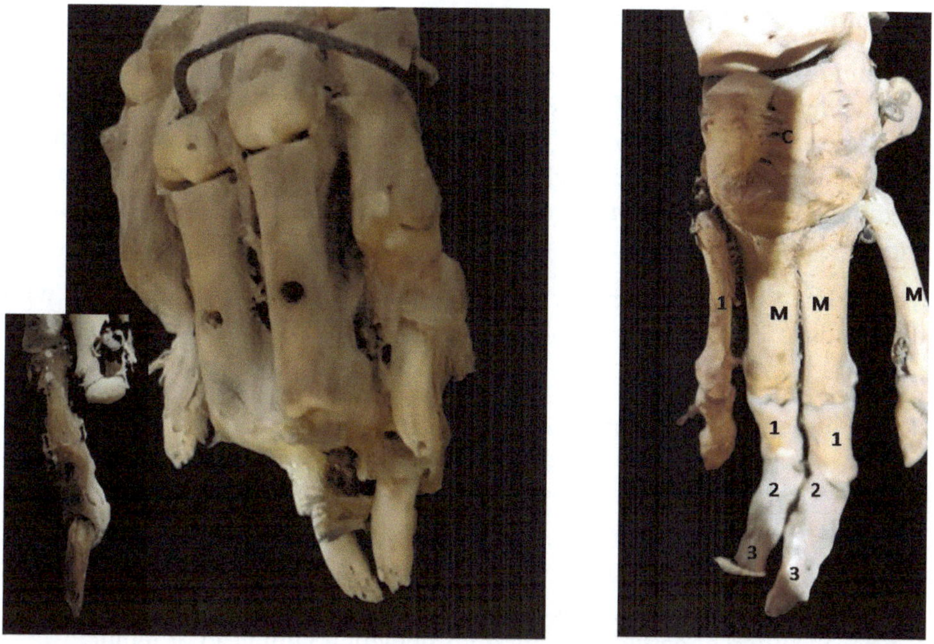

Fig. 3.25: Photograph showing the Manus of A- Dog and B- Pig showing C- Carpals, M-Metacarpals, 1- lst phalanx, 2- 2nd phalanx and 3- 3rd phalanx, arrow showing Crista Unguicularis

B. Comparative Study of the Skeleton of Pelvic Limb

The hind limb comprises four regions viz., hip or haunch, the thigh, the leg and the hind foot or pes. The skeleton of hip is formed by the innominate bone of os-coxae; the thigh contains a large main bone - the femur, and flatting or sesamoid bone termed the patella; the leg contains two bones - the tibia and the fibula; and the pes contains three sets of bones - the tarsus, the metatarsus, and the digits.

The Pelvic Girdle: It consists of the os-coxae, which unites ventrally with the opposite bone at the symphysis pelvis, and articulates with the sacrum dorsally.

a. Comparative Study of Os-Coxae or Innominate Bone

Os-Coxae of Ox

It is the largest flat bone of the body. It is very irregular and is directed upward and backward. In foetal life it consists of three distinct segments - the Ilium, the Ischium and the Pubis, which meet together to form a large articular cavity - the acetabulum or cotyloid cavity for articulation with head of femur.

Ilium

It is triangular flat bone at craniolateral aspect of os-coxae. It presents 2 surfaces, 3 angles and 3 borders. Upper part is flat known as wing and distal angle is narrow considered as shaft (Fig. 3.26). The wing is wider and has two surface viz. external/gluteal and internal/pelvic surfaces. The body is three sided and prismatic in form. It has three surfaces viz. External, internal and ventral surface.

Gluteal surface or External surface: The gluteal surface or external surface is directed outward and backward. Its anterior part is wide and concave while posterior part is narrow and convex. A line, gluteal line runs nearly parallel to the lateral or cotyloid border and crosses the wide part of external surface. It starts from a little below the coxal tuber and continuous with the superior ischiatic spine. This surface gives origin to the *Middle gluteus and the Deep gluteus muscles*. The external surface of the body is convex from before backward and slightly concave from above downward.

The pelvic or internal surface: Faces in the opposite direction of the gluteal surface i.e. ventromedially. It is smooth, convex and consists of two distinct parts which are separated by a rounded ridge. The medial or sacral part is rough and triangular. It presents an "articular facet" for the sacrum. The lateral or the iliac part is quadrilateral and is covered by iliacus muscle. The pelvic surface is crossed by a line which begins below the auricular facet, and is continued on the shaft of bone to join the anterior border of pubis (*Pecton pubis*) and forms the lateral boundary of the pelvic inlet. This line is known as "Ilio- Pectineal Line". Which bears about its middle a tubercle, the Psoas tubercle which gives insertion to *Psoas minor* muscle. The pelvic surface of the body is smooth. The ventral surface present only in the shaft/ body. It is rough and is bounded medially by the psoas tubercle (Fig. 3.26).

Cranial border or the crest of Ilium is concave and thin in the middle and thick and convex towards each end. The medial border or sciatic border is convex and sharp above and concave, thick and rounded over the body. Its middle portion forms the "Greater sciatic notch" which forms the cranial boundary of the greater ischiatic foramen which gives passage to the greater sciatic nerve, gluteal nerves and anterior gluteal vessels. In its lower part it is rounded, rough and is continuous with the superior ischiatic spine. The superior-ischiatic spine gives attachment to the *sacro sciatic ligament* at its free edge and to the *Deep gluteus* muscle on its lateral aspect. Lateral/ cotyloid border is concave and rough lower part it is rounded rough and is continuous with the- superior ischiatic spine. The nutrient foramen is usually situated on the body on its dorso-lateral aspect.

External angle/ lateral angle/ coxal tuber is large thick triangular mass which forms the point of the hip. It is roughened for muscular attachment and gives origin to *Tensor fascia lata mus*cle. Internal / medial angle is blunt called as sacral tuber which articulate with wing of sacrum. It is truncated and is situated opposite the first sacral spine but does not extend as high as the spine itsel£ It is somewhat thickened and rough. Distal or cotyloid / acetabular angle is expanded and join with angles of pubis and ischium to form cotyloid cavity (Fig. 3.26 and 3.28). It forms two-fifth ($2/5^{th}$) part of the acetabulum. Anteriorly it presents supra-acetabular fossa which gives origin to *rectus femoris muscle.* Its prominent dorsal border forms part of superior ischiatic spine which is roughened laterally and smooth medially. The angle is connected with the wing of the bone by a constricted part the shaft (Fig. 3.26).

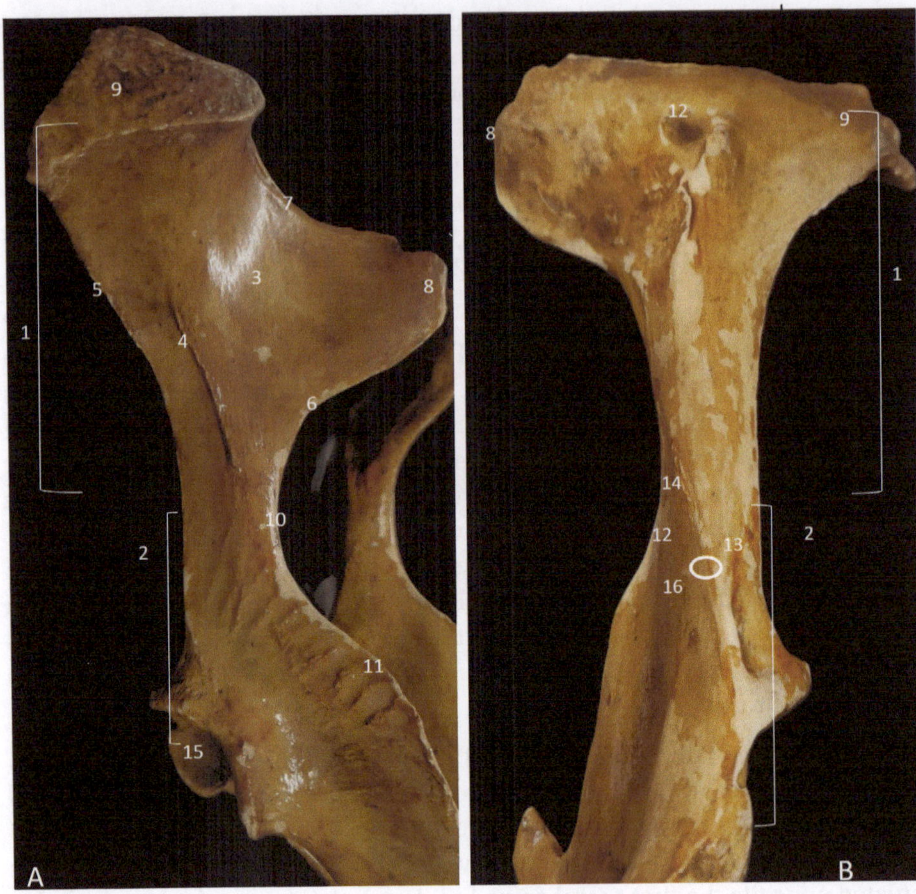

Fig. 3.26: Photograph showing the ilium of Ox A. Gluteal/ External surface, B. Pelvic/ Internal surface

1. Wing of ilium,
2. shaft of ilium,
3. Glureal surface
4. Gluteal line,
5. Lateral border
6. Medial border
7. Dorsal border/ iliac crest
8. Sacral tuber,
9. Cocxal tuber,
10. Greater isciatic notch
11. Greater ischiatic spine
12. Pelvic surface
13. Ventral surface
14. Ilio pectineal line,
15. Acetabulum,
16. Psoas tubercle

Ischium

It is roughly quadrilateral plate situated behind the pubis and form the floor of pelvis. Its long axis is directed obliquely upward and backward, forming an angle of about 50 to 60 degrees with the horizontal plane (Fig. 3.27). The transverse axis is oblique, directed downward and inward forming the similar angle with the horizontal plane so that this part of the pelvic floor is deeply concave from side to side. It presents two surfaces (pelvic and ventral), four borders (cranial, caudal, lateral and medial) and four angles (cranio-lateral, cranio-medial, caudo-lateral and caudo-medial).

The Pelvic surface is deeply concave in both directions and is directed obliquely upward and backwards. It forms the caudal part of the pelvic floor. The Ventral surface is convex in both directions. There is a ridge in its middle for muscular attachment (*M. Biceps Femoris*) and origin of *Adductor muscle.*

Cranial border/ anterior border is thin and concave which forms the caudal/ posterior boundary of obturator foramen. Caudal border is rough and slopes medially and forward to meet the border of the other side forming with it the ischial arch (which forms the inferior boundary of the pelvic outlet). Medial border meets the opposite bone at the symphysis ischii. The symphysis bears a ventral ridge which fades out near the ischial arch. Lateral border is thin and concave. It forms the "Lesser Ischiatic notch" and is continued with the superior ischiatic spine. The notch forms the boundary of the lesser Ischiatic foramen which is for the passage of caudal gluteal vessels.

Cranio-medial angle or symphyseal branch meets the pubis, with which it forms the medial boundary of the obturator foramen. Cranio-lateral angle or acetabular branch joins the other two bones at the acetabulum. It bears a part of ischiatic spine. Caudo-medial angle meets with opposite to form ischial arch. It fuses with opposite side. Caudo lateral angle is expanded and forms the trifid ischial tuberosity. It gives origin to *Biceps femoris, the semimembranous and the semitendinosus muscles* (Fig. 3.27).

The Pubis

It is the smallest and irregularly triangular bone of the three parts of Os-coxae which forms the cranial part of the pelvic floor. It has two surfaces viz. pelvic and ventral surface, three borders viz. cranial, caudal and medial borders and two symphysial and acetabular branches (Fig. 3.27 and 3.28).

The Pelvic surface is concave and smooth form side to side and convex craniocaudally. The urinary bladder rests on it in life. The Ventral surface is convex and rough for muscular attachment.

Cranial border is thin in its middle part (except in young ones and in bulls). Laterally it bears the iliopectineal eminence and which is continuous with the iliopectineal line. A ridge, pectan pubis extends laterally behind the iliopubic eminence and serves as an attachment point for muscles and ligament. The Medial border joins the opposite bone at the symphysis pubis. The caudal border forms the cranial margin of the obturator foramen, and is marked laterally by the obturator groove.

Symphysial branch / cranio-Medial angle form the pubic symphysis. Acetabular branch / lateral angle joins with ilium and ischium to form acetabulum. Caudo-medial angle joins with ischium to form inner boundary of the obturator foramen.

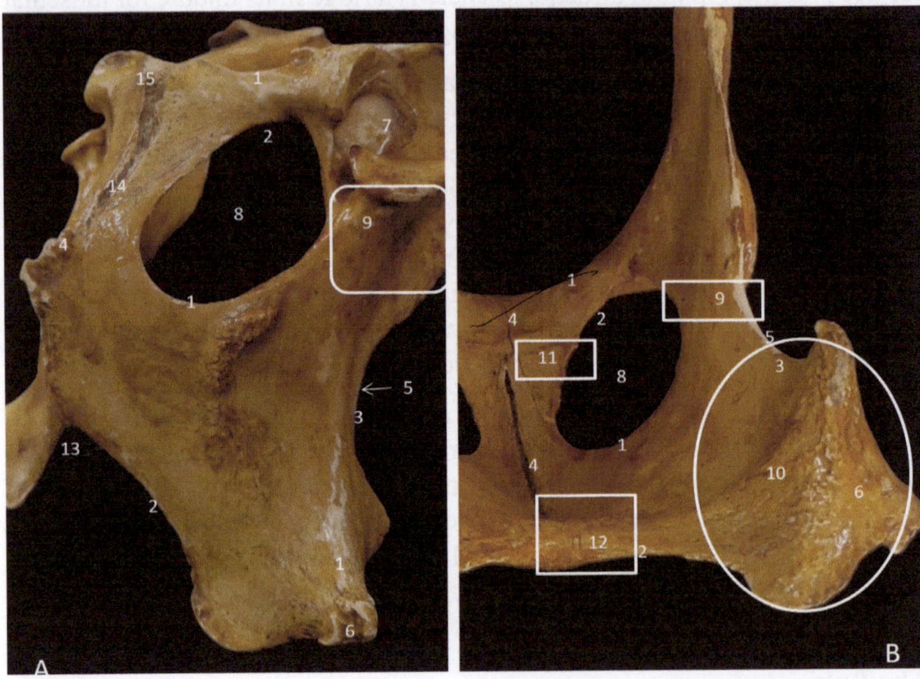

Fig. 3.27: Photograph showing the Ischium and Pubis of ox A. External surface, B. Pelvic/ Internal surface

1. Cranial I border
2. Caudal border
3. Lateral border
4. Medial border
5. Lesser ischiatic notch
6. Ischial tuber
7. Acetabulum
8. Obturator foramen
9. Cranio-lateral angle/ acetabular branch
10. Caudo lateral angle
11. Cranio medial angle/ symphyseal branch
12. Caudo medial angle
13. Ischial arch
14. Ischial symphsis
15. Pubic symphysis

Acetabulum : It is a deep articular depression/ cotyloid cavity formed by the union of corresponding angles of ilium, ischium and pubis where head of femur articulate (Fig. 3.28). It consists of an articular and a non-articular part. The articular part is crescentric, and is cut into internally by a rough non-articular depression, the acetabular fossa. The rim is rounded and is usually marked by two notches. One of these is caudo-medial and is narrow and deep; it leads to the deep acetabular fossa and is often converted into a foramen by a bar of bone. The other notch is cranio-medial, small and sometimes replaced by a foramen or absent (Fig. 3.28).

Obturator foramen: It is the largest, oval shaped foramen of the body formed by the pubis and ischium bones and is largely covered by a fibrous membrane called the obturator membrane. Its axis is directed outward and forward. Obturator membrane and and *obturator muscles* together cover this foramen by leaving a narrow space, obturator canal for passage of obturator blood vessels and nerves (Fig. 3.28).

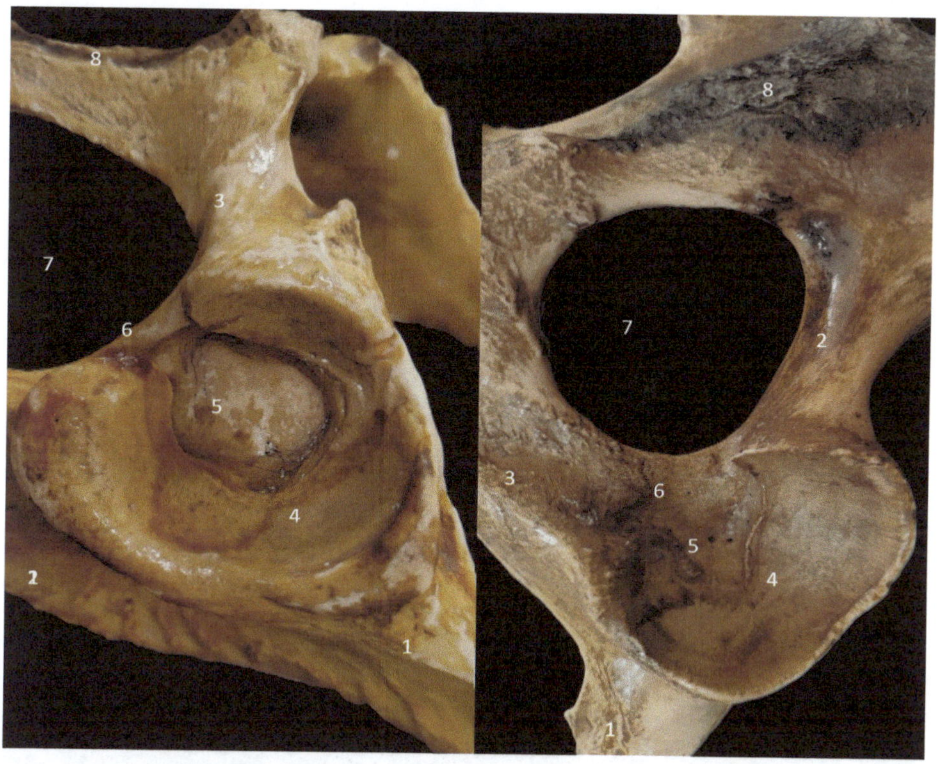

Fig. 3.28: Photograph showing the Acetabulum of A. Ox, B, Horse 1. Ilium, 2. Ischium, 3. Pubis, 4. Articular part, 5. Non articular part/ acetabular fossa, 6. Caudo medial notch, obturator foramen and 8. pelvic symphysis

The Pelvic Cavity or Bony Pelvis: It is a space between two os-coxae. It is the smallest and caudal of three visceral cavities of the body. It is composed of ossa coxarum, the sacrum and the first three coccygeal vertebrae. The dorsal wall or roof is formed by the ventral surface of the body of the sacrum and first three coccygeal vertebrae. It is concave in both directions. The ventral wall or floor is made up by the pubis and ischial bones. It is deeply concave particularly in the transverse direction. The lateral walls are formed by the ilia and the acetabular part of the ischii. In living animal/ fresh state the lateral walls are completed by the *sacro-sciatic ligament and semimembranosus muscle*. The cranial aperture or pelvic inlet is oblique and elliptical in shape. It is bounded by the terminal line or brim; which is composed of the bone of the sacrum dorsally, the lio-pectineal line laterally and the cranial border of the pubis (pecten pubis) ventrally. It has two principal diameters, of these the conjugate or sacro-pubic diameter is measured from the sacral promontory to the cranial most end of the symphysis pubis. The Transverse Diameter is measured at the greater width i.e. just above the psoas tubercles. The Caudal aperture or pelvic outlet is much smaller and is very incomplete in the skeleton. It is bounded dorsally by the third coccygeal vertebra and ventrally by the ischial arch. In the fresh state it is completed by the sacrosciatic ligament and the semimembranosus muscle.

The AXIS: of the pelvis is an imaginary line drawn through the centers of the inlet, cavity and outlet.

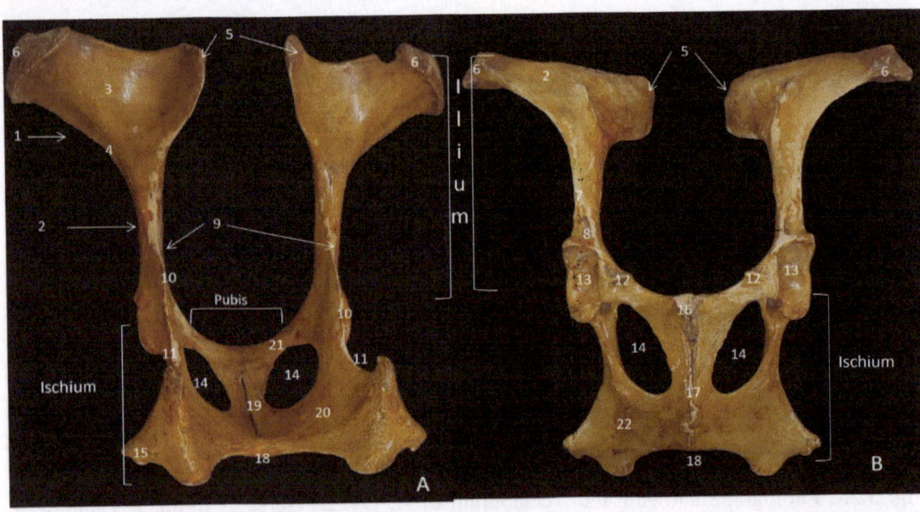

Fig. 3.29: Photograph showing the Pelvic girdle of ox A. dorsal view showing internal/ pelvic surface, B. Ventral view showing external surface

1. Wing of ileum, 2. Shaft of ilium, 3. Glureal surface 4. Gluteal line, 5. Sacral tuber, 6. Cocxal tuber, 7. Ilio pectineal line, 8. Psoas tubercle, 9. Greater ischiatic notch, 10. Ischiatic spine, 11. Lesser ischiatic notch, 12. Pacten pubis, 13. Acetabulum, 14. Obturator foramen, 15. Ischiatic tuber, 16. Pubic symphysis, 17. Ischial symphysis, 18. Ischail arch, 19. Floor of pelvic cavity, 20. Pelvic surface of ischium, 21. Pelvic surface of pubis, 22. External surface of ischium and 23. External surface of pubis

Table 3.8: Comparative Table on Ilium of Ox, Horse, Dog and Pig

Characters	**Ox**	**Horse**	**Dog**	**Pig**
Direction of ilium	Oblique	Oblique	sagittal	Nearly Sagittal
Wing of ilium	Nearly horizontal, irregularly triangular and wide	more extensive, bend laterally	Slightly bend laterally	As ox and horse
Gluteal Surface	Concave, not divided	Same as ox	Same as ox	Divided into two fossae by ridge
Pelvic surface	Has a rounded ridge separating two parts of pelvic surface	rounded ridge is absent	Flat in outline	extensive rough area presents caudally
Gluteal line	A single gluteal line runs parallel to the cotyloid border	Single , runs from middle of the medial border to coxal tuber	Three (cranial, caudal and supra acetabular)	Single line
Iliac crest	Concave and thin	Concave and thin	Strongly Convex and thick and rough	Convex and thick, rough prominent in the middle
Sacral tuber	Truncated, does not rise as high as the sacral spine	Curved dorso- caudally and reach at the level of sacral spine	Two eminence cranial and caudal dorsal iliac spines	Caudally directed, lower than crest
Coxal tuber	Relatively large and prominent Triangular	Quadrangular, having two pairs of tuberosity, more oblique than ox	Two eminence, cranial and caudal ventral iliac spines	Very less thickened
Ilio pectineal line	Less distinct	Same as ox	Very distinct and un interrupted	prominent and thick
Cranial border	Sharp concave iliac crest	Sharp concave iliac crest	Convex and thin	Convex and thin
Body of ilium	Short compressed from side to side having three surfaces- medial, lateral and ventral	Same as ox	Almost sagittal compressed laterally, bears ventro lateral crest	
Greater ischiatic notch	Deep	Same as in ox	Elongated and very shallow	Same as in ox

Table 3.9: Comparative Table on Ischium of Ox, Horse, Dog and Pig

Characters	**Ox**	**Horse**	**Dog**	**Pig**
Pelvic surface	Deeply concave in both directions	Smooth and slightly concave	Same as in ox	Same as in ox
Ventral surface	Convex and ridged	Flat, rough and not ridged	Same as in ox	Presence of crest/ tuberosity
Longitudinal axis	Upward and backward directed	Horizontal	Nearly sagittal	Sagittal in direction
Transverse axis	Obliquely downward and inward	Little slops ventro-medially	Horizontal in axis	Same as in ox
Superior ischiatic spine	High thin and sharp	Low and thick,	Low and thick, having prominent lateral lip	Like ox , but slightly incurved and muscular ridges are more pronounced
Ischial arch	Narrower than horse but deep	Wider than ox but shallow	Smaller than ox and horse and semi elliptical in outline	Very deep and acute
Tuber ischii	Trifid/ tri tuberculated	Not trifid	flattened and everted	Everted and bears three prominences
Ischial symphysis	Bears ventral ridge	Ventral ridge absent	Same as in ox	Same as in ox
Lesser ischiatic notch	Deeper than dog	Same as ox	Shallow and smooth	Similar in size of greater ischiatic notch

Table 3.10: Comparative Table on Pubis of Ox, Horse, Dog and Pig

Characters	Ox	Horse	Dog	Pig
Pelvic surface	Concave side by side Convex- cranio caudally	Convex, concave and smooth in mare	Same as ox	Almost horizontal, Pubis is much thicker in pig than sow
Ventral surface	Convex and rough	Same as ox	Same as ox	Same as ox
Cranial border	Thin bears iliopectineal eminence	Same as ox	Same as ox	Same as ox
Medial border	Forms pubic symphysis	Same as ox	Same as ox	Same as ox
Acetabular angle	Forms acetabulum with ilium and ischium	Same as ox	Same as ox Twice away from coxal tuber than ischial tuber	Same as ox
Obturator foramen	Large, Elliptical shaped	Oval shaped	Equilateral triangular shaped with angles rounded off	Oval / elliptical
Acetabulum	Smaller than horse	Hemispherical	Almost round	Placed little caudal than ox
Acetabular fossa	Crecenteric shaped and deep	Same as in ox	Deep	Deep
Acetabular notch	Two- cranio- medial and caudo-medial	One	One	One
Iliopectineal eminence	Rough	Rough	Nodular thickening	Prominent

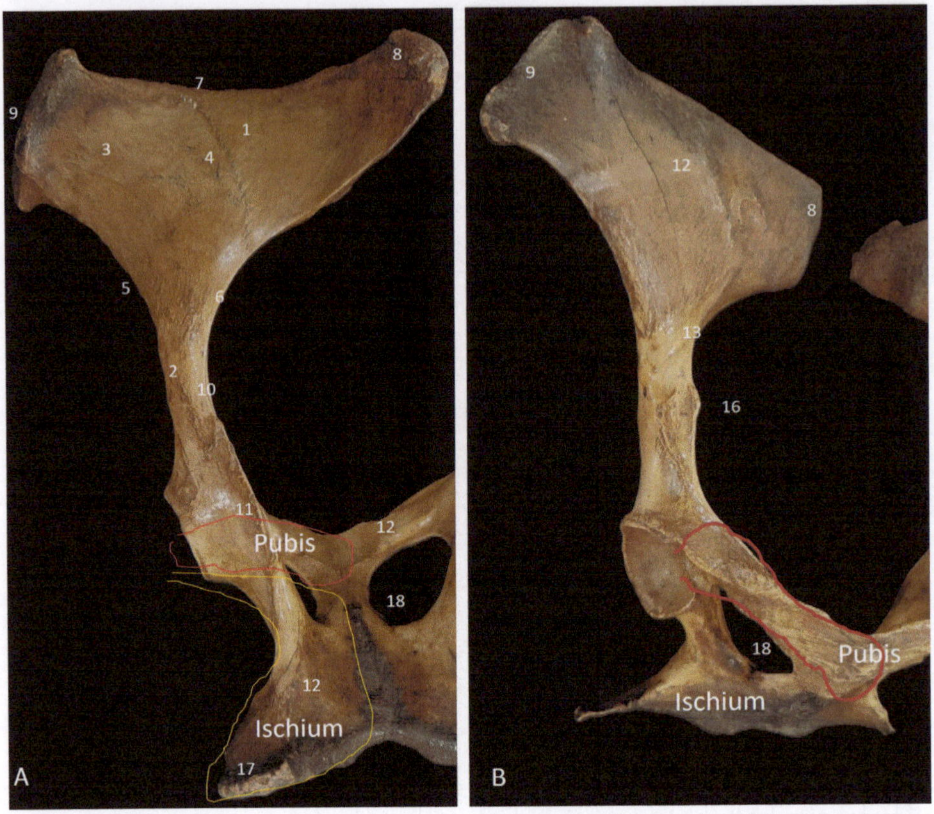

Fig. 3.30: Photograph showing the Oscoxae of Horse- A. External view, B. Internal view Ilium of Horse A. gluteal/ external surface, B. Pelvic/ Internal surface,

1. Wing of ilium
2. shaft of ilium
3. Gluteal surface
4. Gluteal line
5. Lateral border
6. Medial border
7. Dorsal border/ iliac crest
8. Sacral tuber,
9. Coxal tuber,
10. Greater isciatic notch
11. Greater ischiatic spine
12. Pelvic surface 13. Ventral surface
14. Ilio pectineal line,
15. Acetabulum
16. Psoas tubercle
17. Ischial tuberosity
18. Obturator foramen

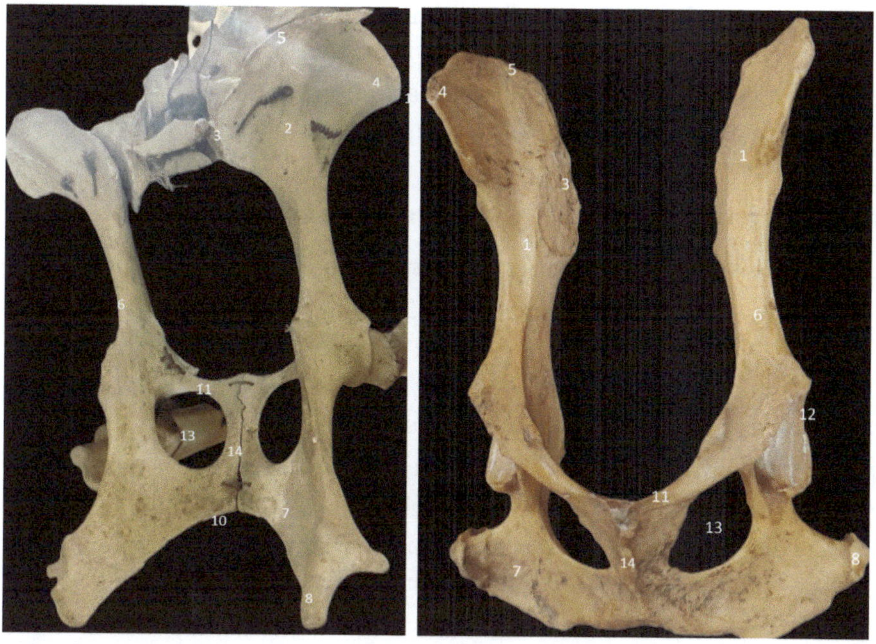

Fig. 3.31: Photograph showing the Pelvic girdle of A. Pig (dorsal view) B. Dog (ventral view)

1. Pelvic surface
2. Gluteal surface
3. Sacral tuber
4. Coxal tuber
5. Dorsal border
6. Shaft of ilium
7. Ischium
8. Ischail tuberosity
9. Lesser ischiatic notch
10. Ischial arch
11. Pubis
12. Acetabulum
13. Obturator foramen
14. Pelvic symphysis

Sexual differences of pelvic girdle of bull and cow

1. The average conjugate diameter is about 9 to 10 inches in the cow and 7.5 inches in bull.
2. The transverse diameter is 7 to 8 inches in cow and about 8 inches in bull.
3. The pelvic outlet is also larger in the cow.
4. The ischial arch being about (1/3) one-third wider than the bull.
5. The cavity of the pelvis is much roomier in the female; diameter between the middles of the superior ischiatic spines is about 8 inches in the cow, 6 inches in the bull.
6. The pubic part of the floor in the female is concave and lies considerably lower than the ischiatic part. In the bull the pubis is very thick medially and this part of the floor is convex.
7. The obturator foramina are correspondingly larger in the female.

8. The ilium is shorter, especially with regards to its shaft, and the greater ischiatic notch is deeper and narrower in male.
9. The pelvis of the bullock, when the castration has been performed early, resembles to that of cow, otherwise the male characters appear to be retained to a large degree.

Sexual differences of pelvic girdle of stallion and mare

1. The inlet is almost circular in mare and semi-elliptical in stallion.
2. The average conjugate diameter is 9.5 inches in the mare and 7.5 inches in the stallion.
3. The transverse diameter is about the same as the conjugate diameter in the mare but is about 8 inches in the stallion.
4. The outlet of the pelvis is also larger in the mare. The ischial arch is about one-third (1/3rd) wide than in the stallion.
5. The cavity is much more roomy in mare. Diameter between the middle of the ischiatic spines is about 8 inches in the mare and about 6 inches in the stallion.
6. The pubic part of the floor in the female is concave and lies considerably lower than the ischiatic part, which is wide, and almost flat.
7. In the stallion the pubis is very thick medially, and this part of the floor is convex, while the ischial part is relatively narrow, and is concave from side to side.
8. The obturator foramina are correspondingly larger in the female

The Pelvis of Dog: The inlet of the pelvis is very oblique. It is almost circular in the female, but in the male it is elliptical and the conjugate diameter is longer. The cavity is narrowest between the acetabulum, and very wide behind. The floor is concave and relatively narrow infront, wide and flat behind.

The pelvis of pig: Pelvic inlet is Elliptical and very oblique while outlet is Inclined ventrally. Diameter of pelvic outlet is 7.5-8.75 cm. Conjugate diameter is 11-12.5 cm in male and 12.5-15 cm in female. Transverse diameter is about 8.75-10cm in sow. Ischial arch is narrower and deeper in male and wide and shallow in female. Pubic floor is Concave from side to side less sloppy than sow. The obturator foramina are correspondingly larger in the female

b. Comparative Study of Femur and Patela

Femur of Ox

Femur is the largest and strongest cylindrical bone of hind limb which is directed obliquely downward and forward. It articulates above with os-coxae bone forming hip joint and below with tibia - fibula and patella bones forming stifle joint. It consists of a shaft/ body having four surfaces and two, proximal and distal extremities.

The shaft is cylindrical upto middle and three sided below. It has four surfaces viz. cranial, Caudal, medial and lateral (Fig. 3.33). The lateral, medial and cranial surfaces are continuous and convex from side to side. The cranial surface is confined to the proximal third of the body; lower down a rounded ridge separates the medial and lateral surface. All these surfaces except the caudal one are covered by *Quadriceps femoris muscle*. The caudal surface is strong and bears ridges for attachment of *Adductor muscles* and is slightly concave from above downward. It is wider over its distal third than on the proximal and middle thin and is rough and wide below (Fig. 3.33).

The medial border bears on its proximal part trochanter minor, a rough tuberosity which encroaches on the caudal surface and to which *Ilio psoas muscle* is attached. A rough line curves from trochanter minor upto the front of the neck indicates the caudal limit of attachment of the *Vastus medialis* muscle. On lateral border near the distal extremity a rough shallow depression, the supracondyloid fossa is situated from which *Superficial digital flexor muscle* arises. The fossa is bounded laterally and medially by lateral and medial supra-condyloid crest respectively. Lateral and medial head of *Gastrocnemius muscle* arises from this lateral and medial supracondyloid crests, respectively. On medial border between the trochanter minor and medial supracondyloid crest the area is covered by *Pectineus muscle.*

Proximal extremity comprised of head, neck and greater trochanter. The head is situated medially and projects upward inwards. Its articular surface which is hemispherical extends considerably on the upper surface of the neck and articulates with the acetabulum. Medially a small non articular depression termed as "fovea capitis" is present, which furnishes attachment to the round ligament of hip joints. The neck is constricted part below the head. Greater trochanter is undivided, convex and rough and is situated laterally. It rises fairly high above the level of head and gives attachment to the *Deep gluteus muscle.* Its external surface is rough and in fresh state is coated with cartilage over which the tendon of *middle gluteus muscle* passes to be inserted into the crest lower down. A ridge, known as trochanteric ridge runs obliquely and connect the greater and lesser trochanter. Trochanteric ridge bounded deep trochanteric fossa which encloses the depression between ridge and head. The trochanteric ridge is for the lodgment of *gluteus medius* and trochanteric fossa is for *gemellus, obturator externus* and *obturator internus muscles* (Fig. 3.32).

Distal extremity comprises of the trochlea in-front and condyles behind. The trochlea comprises of two unequal ridges out of which the medial ridge is much wider, more prominent, and extends upward. The ridges are separated by a groove and form an extensive surface for articulation with the patella. The condyles are situated behind and articulate with the condyles of tibia and the menisci (Fig. 3.32). The medial and lateral condyles are separated by inter-condyloid fossa which lodges spine (Inter condylar eminence) of the tibia and the cruciate ligaments of the stifle joint. Between the lateral condyle and the lateral ridge of the trochlea a small, shallow extensor fossa is present from where common tendon of *Long digital extensor* and *Peroneus Tertius* muscle is originated. On each side of condyles epicondyles are present. The medial epicondyle is a rounded prominence on the medial surface of the distal extremity, to which the collateral ligaments are attached. The corresponding lateral epicondyle is less salient. It presents a mark where lateral ligament is attached, below and behind this there is a depression in which the *Popliteus muscle* arises (Fig. 3.34).

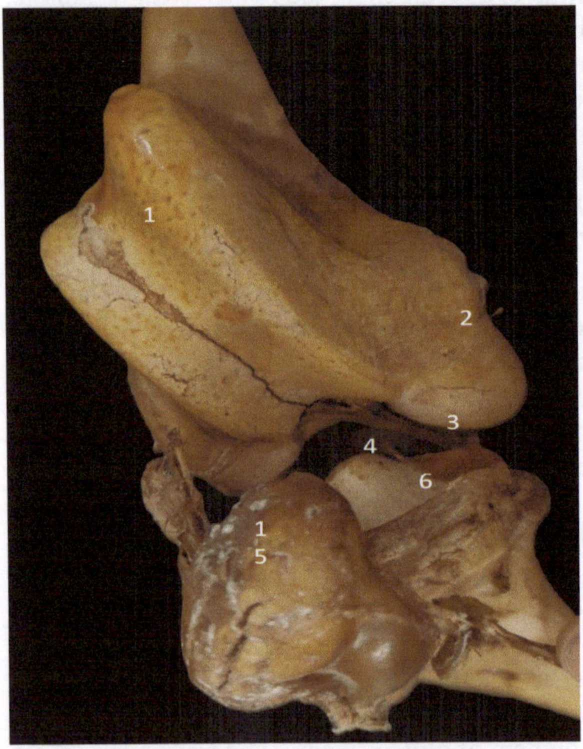

Fig. 3.33: Photograph showng the Stifle joint of ox

1. Trochlea of femur, 2. Condyles of femur, 3. Condyles of tibia, 4. Tibial spine, 5. Patella
6. Menisci

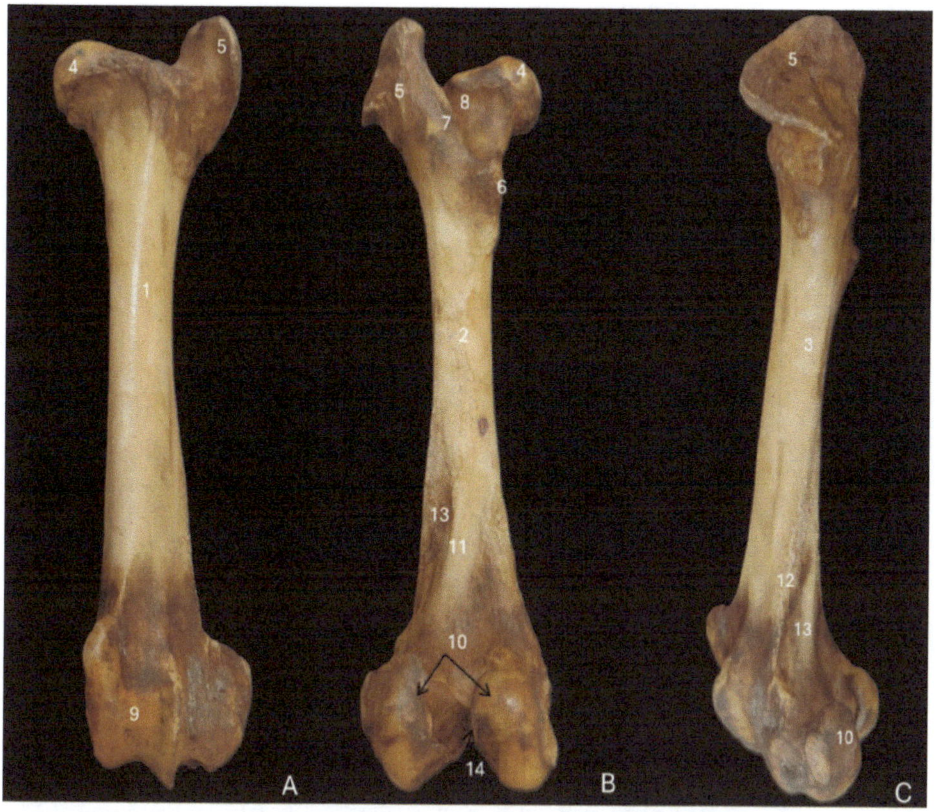

Fig. 3.34: Photograph Showing the Left Femur of Ox A- Cranial view, B. Caudal view, C. Lateral view

1. Cranial surface, 2. Caudal surface, 3. Lateral surface, 4. Head, 5. Trochanter major, 6. Trochanter minor, 7. Trochanteric ridge, 8. Trochenteric fossa, 9. Trochlea, 10. Condyles, 11. Supra condyloid crest (medial), 12. Supra condyloid crest (lateral), 13. Supra condyloid fossa, 14. Intercodyloid fossa

Table 3.11: Comparative Table on Femur of Ox, Horse, Dog and Pig

Characters	Ox	Horse	Dog	Pig
Natural position	Directed obliquely downward and forward	Same as in ox	Same as in ox Relatively much longer than in the horse, and ox.	Same as in ox
Shaft	Cylindrical upto middle and three sided below	Cylindrical, Larger above than below	Regularly cylindrical except near extremities, where it is wider and compressed cranio-caudally	Wide and massive, and is distinctly quadrilateral
Head	Hemispherical in shape	Same as in ox	little more than a hemisphere	More curved than ox
Fovea capitis	Smaller than horse, notch is absent and gives attachment to round ligament only	Deeply notched and is larger than in the ox. the accessory and round ligaments are attached.	Shallow fovea behind and lateral to its centre	Instead of this large tuberosity is present towards medial side
Trochanter major	Undivided	Divided into two parts by a notch. The cranial anterior part , convexity and caudal part raised above to form summit.	Does not extent as high as the head	Does not extend above the head
Trochanter minor	Form of a rough tuberosity	form of a thick rough ridge and situated lower down than in the ox	Blunt tuberosity	Similar to ox
Trochanter tertius	Absent	Present on lateral border, curved forward and has a thick edge	Very small lateral eminence of the proximal body	Absent
Trochanteric fossa	Deep	Shallower than ox	Round and deep	Same as in ox
Trochanteric ridge	Obliquely directed	Straight	Obliquely directed	Same as in ox
Supra condyloid crest	Two crests, the medial one being smaller tan lateral	More prominent than ox	Two rough lines present on caudal surface which diverge towards extremities	Medial crest is indistinct , lateral one is in the form of narrow ridge
Supracondyloid fossa	Shallower than horse	larger and deeper than ox	Not present	Not present

Characters	Ox	Horse	Dog	Pig
Supra Condyloid tuberosities	Not present	Not present	Two in number, medial one is smaller than lateral one	Lateral one is larger than medial
Ridges of the trochlea	Unequal, medial one is much wider, more prominent, and extends upward	More oblique than ox	sagittal in direction and are almost similar	sagittal in direction and are almost similar
Facets on Distal condyles	Absent	Not present	Facets are present on the caudal side of distal condyles	Absent
Inter Condyloid fossa	Shallower than horse	Deeper than ox	Wide	Deep

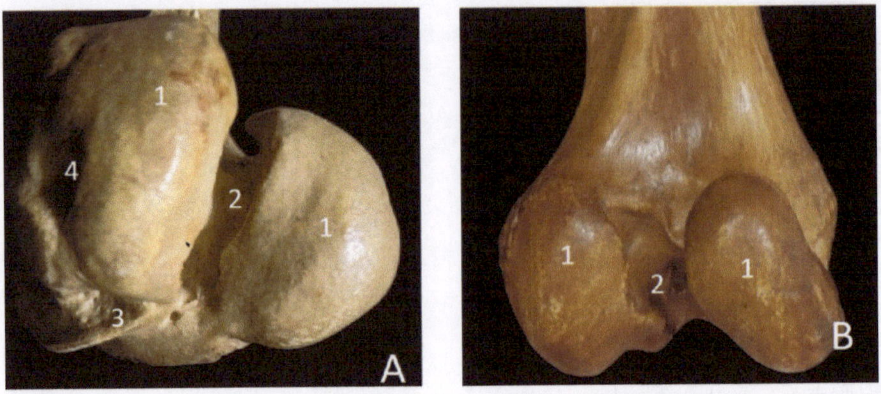

Fig. 3.34: Photograph showing the Caudal view of Distal extremity of left Femur of A. Ox and B. Horse

1. Condyles
2. Inter-codyloid fossa
3. Extensor fossa
4. Fossa for popliteal muscle

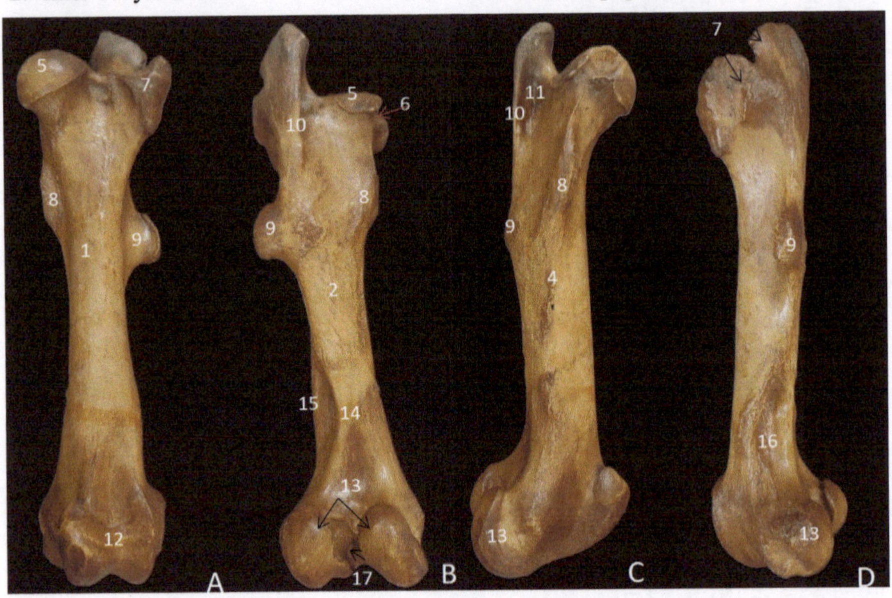

Fig. 3.35: Photograph showing the Left Femur of Horse A. Cranial view, B. Caudal view, C. Medial view and D. Lateral view

1. Cranial surface
2. Caudal surface
3. Lateral surface
4. Medial surface Head
5. Fovea capitis in head
6. Trochanter major
7. Trochanter minor
8. Trochanter tertius
9. Trochanteric ridge
10. Trochanteric fossa
11. Trochlea
12. Condyles
13. Supra condyloid crest (medial)
14. Supra condyloid crest (lateral)
15. Supra condyloid fossa
16. Inter-codyloid fossa

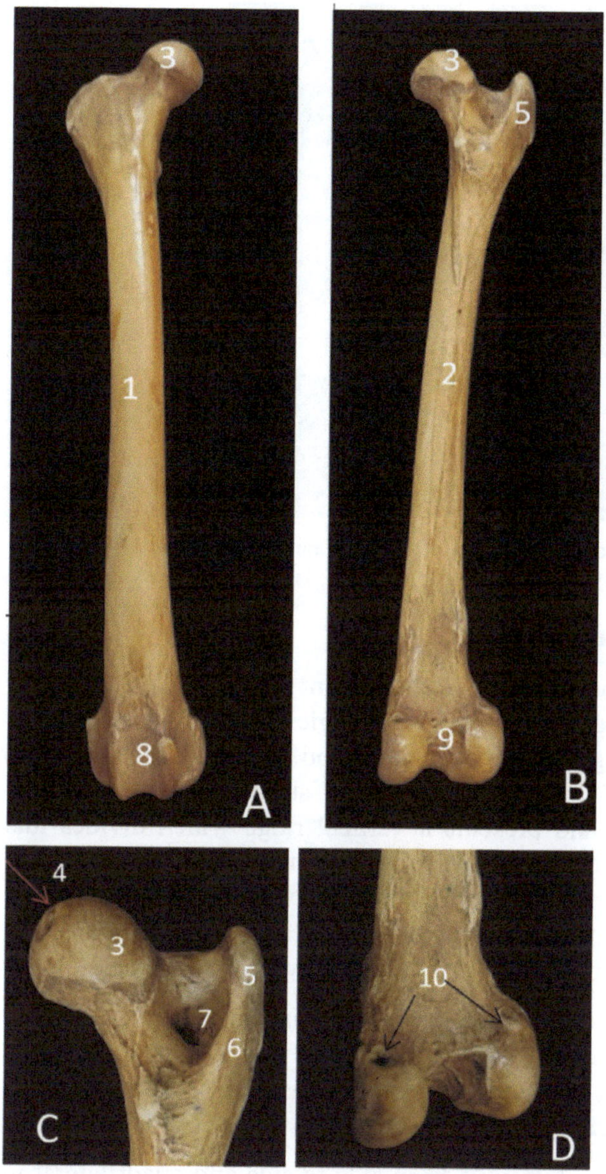

Fig. 3.36: Photograph showing the Right Femur of Dog A. cranial view, B. Caudal view C. Proximal extremity, D. Distal extremity

1. Cranial surface
2. Caudal surface
3. Head
4. Fovea capitis
5. Trochanter major
6. Trochanteric ridge
7. Trochanteric fossa
8. Trochlea Condyles
9. Condyle
10. Facet for fabella

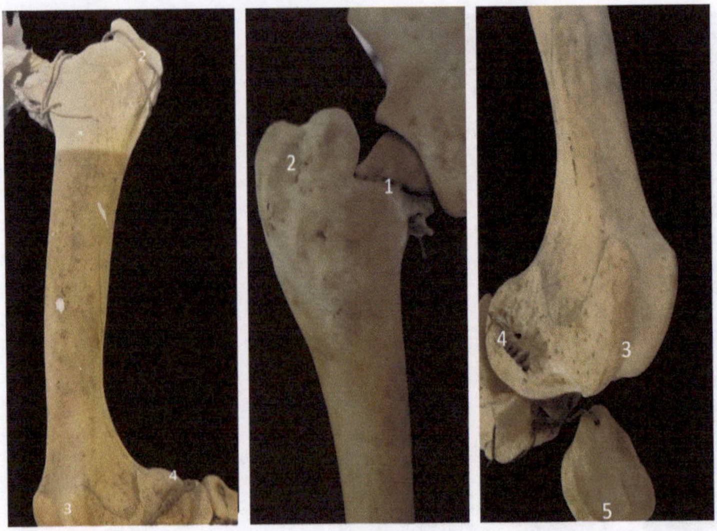

Fig. 3.37: Photograph showing the Left Femur of Pig A. Cranio-lateral view, B. Proximal extremity C. Distal extremity; 1. Head, 2. Trochanter major, 3., Trochlea, 4. Condyles 5. Patella

Comparative Study of Patella of Ox

It is triangular large sesamoid bone placed in front of trochlea of femur. It presents two surfaces, three borders and three angles. The cranial surface is irregularly triangular, strongly convex, irregular and very rough for ligament and muscular attachment (Fig. 3.38). The caudal surface is also triangular but articular. It is smooth and presents a vertical ridge which divides this surface into two concave articular areas for corresponding groove of trochlea. The medial articular area is much larger, and is not very well adapted to the corresponding ridge of the trochlea. However, in fresh state, it is completed and rendered more congruent by the curved accessory fibro-cartilage.

The medial border is concave while the lateral border is rounded. Dorsal border is convex and faces upward and backward to form the base of the patella. Medial angle is larger than the lateral angle which is blunt. Ventral angle is pointed called as apex which is directed downward.

Patella of Horse: It is short, but wide from side to side and quadrilateral in shape. The medial area of the articular surface is more extensive. Medial and lateral borders meet at the base and form wide angle. Medial and lateral angles are not prominent. The base is larger than that of the ox and the apex is blunt pointed (Fig. 3.38).

Patella of Dog: It is comparatively larger than ox and horse. It is short, but wide from side to side and quadrilateral in outline. Its cranial surface is convex.

Caudally the articular surface is concave along the length and convex across the width. The medial area of the articular surface is more extensive. The base is larger than that of the ox and the apex is blunt pointed (Fig. 3.38).

Patella of Pig: It is smaller in size. It has three surfaces (cranial, caudal and compressed transverse surface).

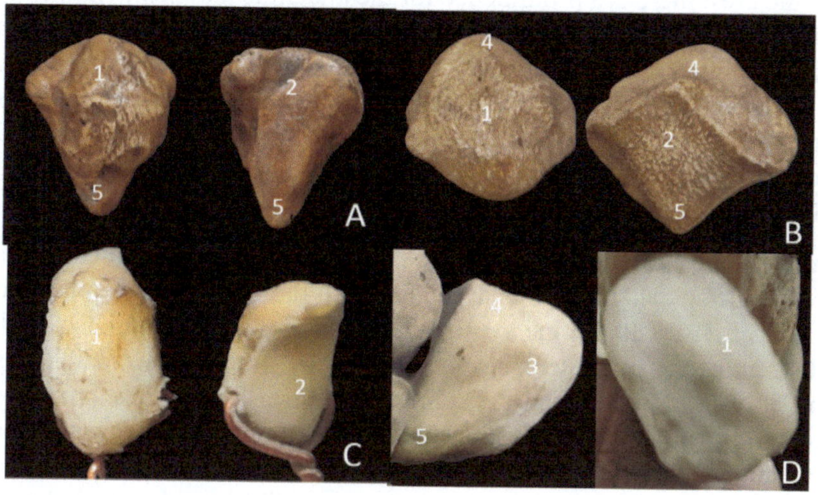

Fig. 3.38: Photograph showing the patella of A. ox, B. Horse, C. Dog and D. Pig-1, Cranial view, 2. Caudal view, 3. Lateral view, 4. Base, 5. Apex

c. Comparative Study of Tibia and Fibula

Tibia and Fibula of Ox

It is strong and massive long bone of hind limb which forms the leg region of the body. It extends obliquely downward and backward (distally –caudally). It articulates above with the condyles of femur to form stifle joint and below with the tarsus to form hock joint, and laterally with the fibula to form tibio-fibular articulation. It comprised of a body and two extremities.

The Shaft is expanded and prismatic in its upper part, twisted, narrow and flat at lower end. It presents three surfaces (lateral, medial and caudal) and three borders (lateral, medial and cranial) (Fig. 3.39). The Lateral surface is smooth, faintly concave and gradually spiral downward to front. It is covered by the *Tibialis anticus muscle*. Medial surface is wide above and narrow below. It has rough prominences for the attachment of the medial ligament of stifle joint and the *Sartorius and Gracilis muscles*; below this it is narrow, convex from edge to edge and sub cutaneous. Caudal surface is rough and presents many lines called *linea musculares* / muscular lines to which the *Deep digital flexar muscles* are attached. On the proximal one fourth area near medial border a narrow triangular area, popliteal line is present which gives attachment to *Popliteus muscle* (Fig. 3.39).

Cranial border presents tibial crest in its proximal third. The crest is in the form of a curved rough line which gradually fades out distally. A little below the middle of the crest there is a prominence for the attachment of *Semitendinosus muscle*. Lateral border is concave in the upper part. It presents interosseous space with the attachment of fibrous cord of fibula during embryonic life. Nutrient foramen is situated on the lateral border at the junction of the proximal and middle third. The medial border is straight and rounded in its proximal third, and curves further downwards. The *Popliteus muscle* is attached on the proximal third of this border (Fig. 3.39).

Proximal extremity is expanded and three sided. It consists of two condyles which are saddle shaped structures articulate with the condyles of femur and meniscus (Fig. 3.39). Tibial spine is an intercondyloid eminence which is bifid. The medial eminence is higher than the lateral one. Intercondyloid fossa is located on, infront of and behind the spine which furnishes the attachment of cranial cruciate ligament and menisci. Caudally the popliteal notch separates the medial and lateral condyles. A tubercle is present on the medial side of the notch for attachment of the caudal cruciate ligament. The lateral condyle bears a facet on its lateral aspect for fusion of the rudimentary head of the fibula. The large cranial eminence, the tibial tuberosity is blunt and situated above tibial crest. It gives attachment to the patellar ligament. Tibial tuberosity is separated from lateral condyle by *sulcus extensorius/ sulcus muscularis* from which *common tendon of origin of the extensor digitalis longus and peroneus tertius passes* (originated from extensor fossa of femur) (Fig. 3.39).

Distal extremity is quadrangular in form and wider/ larger medially than laterally. It presents two sagittal grooves (cochlea) separated by a ridge corresponding to the trochlea of tibial tarsal bone. There is an extensive synovial fossa extending into both the grooves. The lateral groove is wider and shallower than the medial one and articulates with lateral malleolus. The grooves are bounded on either side by malleoli. The collateral ligaments of hock joint attaches with the malleoli. The cranial part of the medial malleolus is prolonged downwards and has pointed end. Behind and above the medial malleolus is a well-marked groove for the tendon of *Flexor digitalis longus* muscle. On the lateral side of the extremity is also a variable groove, for the passage of the *lateral extensor tendon*. On lateral side, a sharp ridge separates the articular surface for tibial tarsal (lateral groove) from the articular facet for lateral malleolus which is a separate piece of fibula bone.

Fibula:-It is highly rudimentary. The proximal end (head) is fused with the lateral condyle of tibia. Body is in the form of blunt prolongation about 1 and ½ inch from head. The distal extremity is in the form of separate piece of bone, lateral malleolus. It is quadrilateral in outline and compressed from side to side which project beyond the cochlea of tibia. Its proximal surface articulates with the distal end of tibia. Proximal border has three processes directed towards tibia. The cranial and caudal processes are blunt and middle one bears a small spine which fits into the groove on tibia. The distal surface rests on the fibular tarsal, and the medial articulates with the lateral ridge of the tibial tarsal bone. The lateral surface is rough and irregular.

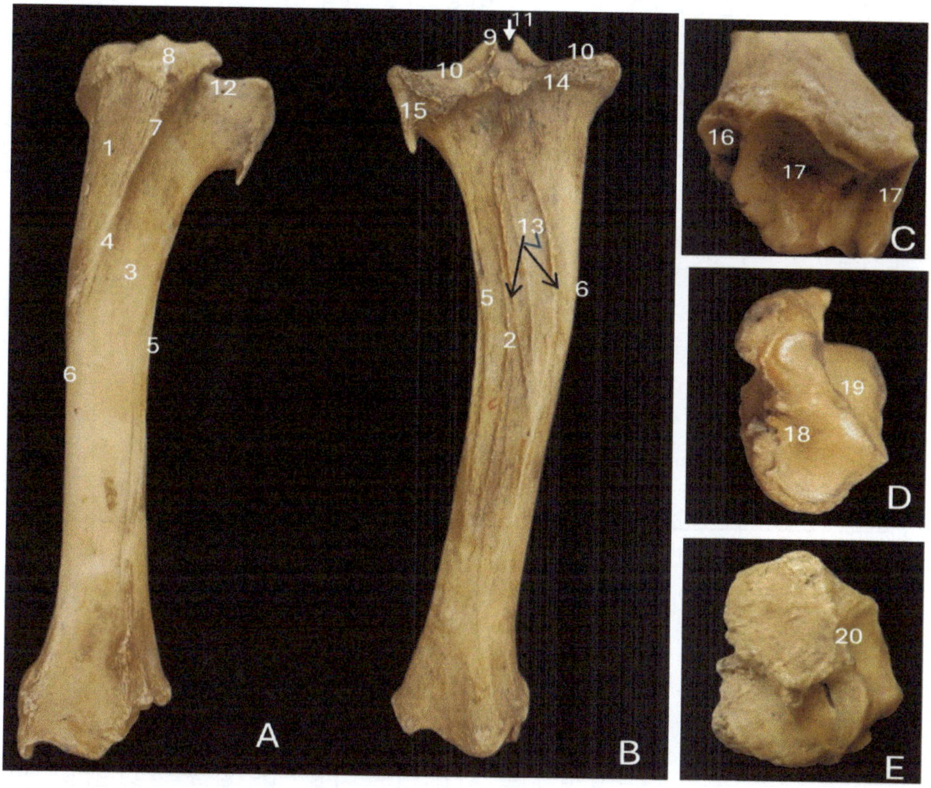

Fig. 3.39: Photograph Showing the Left tibia and fibula of Ox ,A. cranial view, B. Caudal view, C. Proximal extremity cranial view), D. Proximal extremity (caudal view), E. Distal extremity, F. Lateral malleolus (proximal view), G. Lateral malleolus (lateral view)

1. Cranial surface, 2. Caudal surface, 3. Lateral surface, 4. Cranial border, 5. Lateral border 6. Medial border, 7. Tibial crest, 8. Tibial tuberosity, 9. Tibial spine, 10. Tibial condyles, 11. Inter condyloid fossa, 12. Sulcus muscularis, 13. Linea muscularis, 14. Popliteal notch, 15. Fibula, 16. Facet for lateral malleolus, 17. Facet for tarsals, 18. Facet for tibia, 19. Facet for fibuiar tarsal, 20. Facet for tibial tarsal

Table 3.12: Comparative Table on Tibia of Ox, Horse, Dog and Pig

Characters	Ox	Horse	Dog	Pig
Shaft	larger and straight	relatively larger and straight	forms a double curve, the proximal part in convex medially, the distal part laterally	Slightly curved, convex medially
Tibial spine/ eminence	medial eminence is higher than the lateral one	Same as ox	Both medial and lateral tibial spines are at same level	lateral tibial eminence is higher than the medial one
Condyles	Lateral condyle is more projected than medial one , articular surfaces are at same plane	Articular surface of lateral condyle is higher than medial condyle	Both articular surfaces are at same level	Lateral condyle is broader than medial condyle
Fibular articular facet	Fibula fused with articular surface	Same as in ox, fibula articulates with this facet	Present below and behind the lateral condyle	Present caudo-laterally to the lateral condyle
Tibial crest	Prominent	Same as in ox	short and prominent with impression	long and distinct, curved laterally
Tibial tuberosity	Not grooved	Grooved	not grooved but bears a distinct mark	Grooved cranially
Sulcus extensorius	Narrow and shallow	Wider and deeper than ox	Deeper than ox	Deeper and narrower than ox
Popliteal line	Narrow triangular area on the proximal fourth of caudal surface towards medial border	More prominent , runs obliquely from the proximal part of lateral border to the middle of medial border	Absent	Prominent ridge form
Popliteal notch	Very shallow	Deep	Same as in ox	Same as in ox
Linea muscularis	Indistinct	Prominent, faded distally	Less prominent than ox	Salient
Nutrient foramen	Situated on the lateral border	Situated on or near the popliteal line.	proximal third of the lateral border	Same as ox

Characters	Ox	Horse	Dog	Pig
Medial malleolus	pointed process which project distally and participate in tibial cochlea	Tuberculated, participate in tibial cochlea	prominent and project distally	participate in tibial cochlea
Articular grooves of distal extremity	Sagittally directed	Obliquely directed	Almost sagittal	Same as in ox relatively narrower transversaly
Distal articular facet	Articulates with Tibial tarsal, lateral malleolus of fibula	Articulates with Tibial tarsal only, lateral malleoulus is fused	Articulates with Tibial tarsal	Tibial tarsal and lateral; malleolus through incisura

Table 3.13: Comparative Table on Fibula of Ox, Horse, Dog and Pig

Features	Ox	Horse	Dog	Pig
Extent of fibula	Present in the form of small prolongation	Pointed distally and extend upto the middle of the lateral surface of the tibia	extends along the whole length of tibia	extends along the lateral border of tibia
Shaft	In the form of fibrous cord like structure which ossified later on	Only proximal part present which tapers distally	Well developed	Well developed, flattened from side to side
Proximal extremity	fused with lateral condyle of tibia,	Head is large, flat oval plate which articulate with lateral condyle of tibia	Flattened	Flattened and grooved laterally
Distal extremity (Lateral malleolus)	Separate piece of bone (Lateral malleolus). Articulate with tibia (proximally, tibial and fibular tarsal bones	Wider than medial malleolus, fused with the tibia, articulates with tibial tarsal bone	Tuberous and projects distally than tibia, articulate with tibia and the tibial and fibular tarsal bones	Four sided plate shaped, grooved laterally, articulates with tibia, talus and calcaneus
Tuberosities of lateral malleolus	Absent	Absent	two tuberosity with groove	Absent
Interosseous space between tibia and fibula	narrow	Wide	proximal interosseous space distal part is closely applied to the tibia	Wide

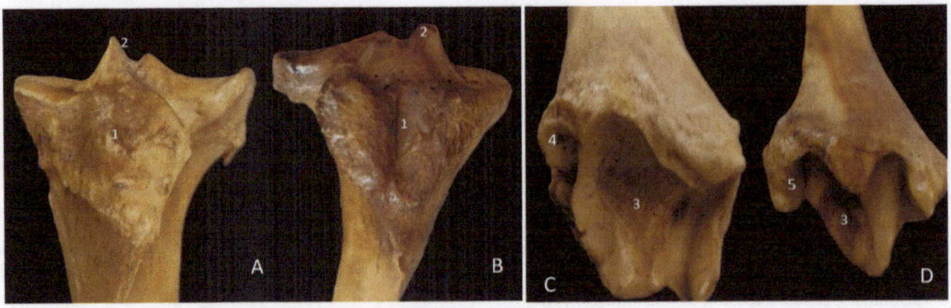

Fig. 3.40: Photograph showing the Proximal extremity of tibia of Ox (A) and Horse (B) and Distal extremity of tibia of Ox (C) and Horse (D): 1. Tibial tuberosity, 2. Medial Tibial spine, 3. Facet for tibial tarsal bone, 4. Facet for lateral malleolus, 5. Lateral malleolus

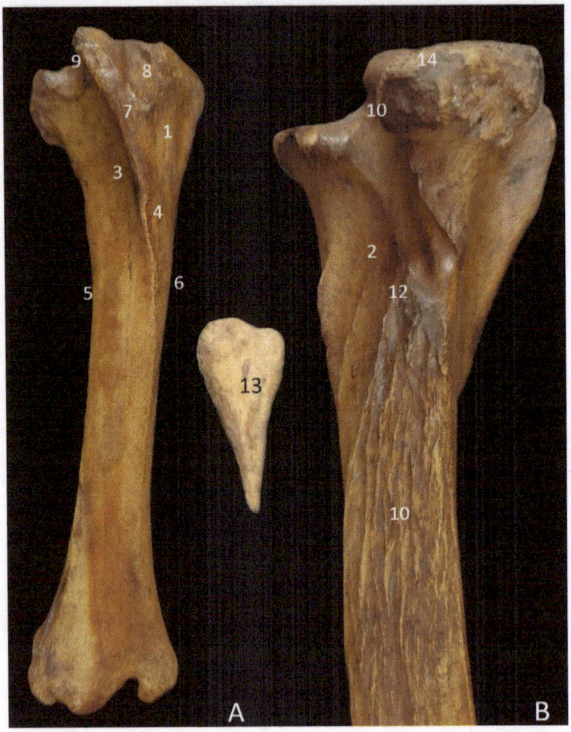

Fig/3.41: Photograph Showing the Right tibia and fibula of Horse A. Cranial view, B. Caudal view

1. Cranial surface
2. Caudal surface
3. Lateral surface
4. Cranial border
5. Lateral border
6. Medial border
7. Tibial crest
8. Tibial tuberosity
9. Sulcus muscularis
10. Linea muscularis
11. Popliteal notch
12. Popliteal line
13. Fibula
14. Facet for fibula

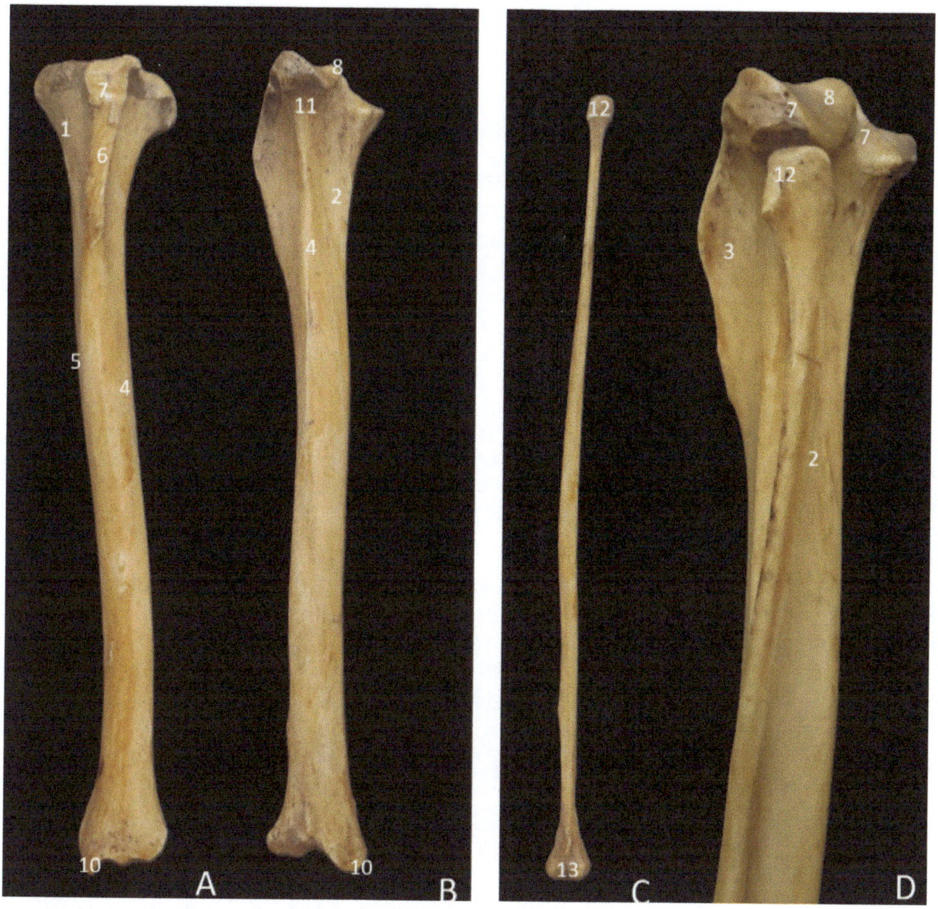

Fig, 3.42: Photograph showing the Left Tibia and Fibula of Dog A. Cranial view, B. Caudo-lateral view, c. Fibula, d. Attachment of fibula

1. Cranial surface
2. Caudal surface
3. Lateral surface
4. Lateral border
5. Medial border
6. Tibial crest
7. Tibial tuberosity
8. Tibial spine
9. Tibial condyles
10. Medial malleolus
11. Facet for fibula
12. Proximal extremity of Fibula
13. lateral malleolus

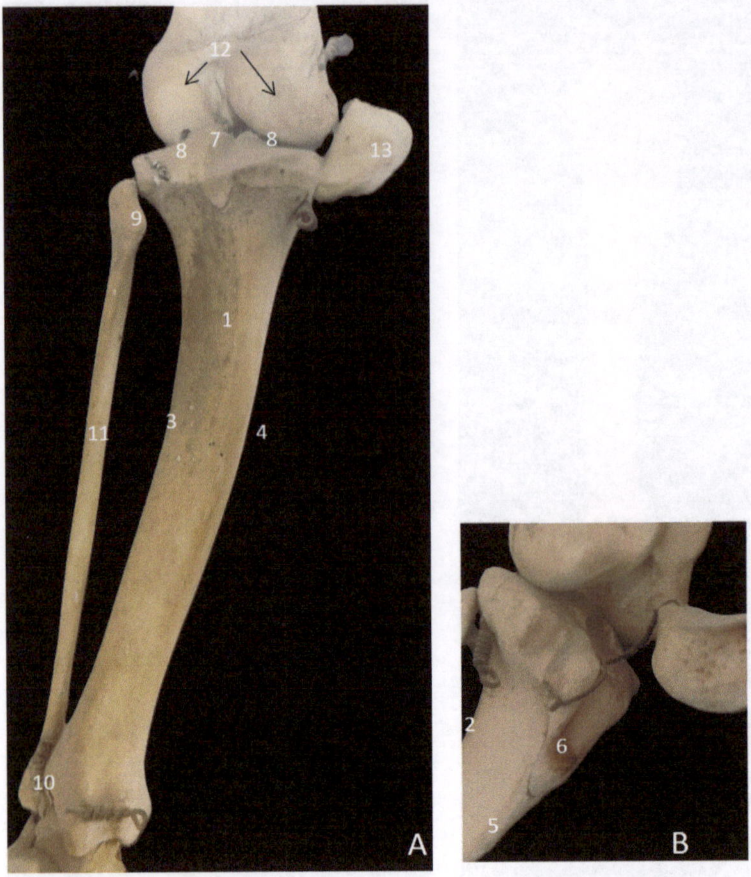

Fig. 3.43: Photograph showing the Left Tibia and Fibula of Pig A. ew, B. Cranial view of Proximal extremity

1. Caudal surface
2. Lateral surface
3. Lateral border
4. Medial border
5. Tibial crest
6. Tibial tuberosity
7. Tibial spine
8. Tibial condyles
9. Proximal extremity of Fibula
10. Distal extremity of Fibula
11. Shaft of fibula
12. Condyles of femur
13. Patella

d. Comparative Study of Tarsals

Tarsal bones/ Tarsus :-The tarsal are short bones of hind limb. Total seven tarsal are present in the body. But their number varies with the species because of fusion of tarsus (Table 3.14). Tarsus are arranged in three rows viz. proximal, middle and distal row. In ox, there are five tarsal bones arranged in different rows.

Table 3.14: Showing the arrangement of tarsals in different species

Species	Rows			
Ox (5)	Proximal row	Talus	Calcaneus	
	Middle row	Central and 4th fused tarsal		
	Distal row	1st tarsal	2nd and 3rd fused tarsal	
Horse (6)	Proximal row	Talus	Calcaneus	
	Middle row	Central tarsal		4th tarsal
	Distal row	1st and 2nd fused	3rd tarsal	
Dog (7)	Proximal row	Talus	Calcaneus	
	Middle row	Central tarsal	4th tarsal	
	Distal row	1st tarsal	2nd tarsal	3rd tarsal
Pig (7)	Proximal row	Talus	Calcaneus	
	Middle row	Central		4th tarsal
	Distal row	1st tarsal	2nd tarsal	3rd tarsal

Tarsus of Ox

Talus (Tibial tarsal):- It is elongated pulley shaped bone situated in a vertical manner on the medial side of hock (Fig. 3.45). It articulates with calcaneus bone which is situated caudolaterally to this. Proximal and dorsal surfaces are continuous and forms a trochlea containing lateral and medial ridges for articulation with distal extremity of tibia. Lateral ridge is thicker than medial ridge and beside the tibia also articulate with lateral malleolus of fibula. Distal surface also contain trochlea having two condyles to articulate with central and 4th fused tarsal bone. It presents a deep fossa on lower part of dorsal surface while caudal surface is smooth and articular having large facet to articulate with calcaneus. Medial surface is flat and rough. Lateral surface is irregularly concave and having two articular facets for calcaneus (Fig. 3.45).

Calcaneus (Fibular tarsal):- It is elongated short bone situated lateral and caudal aspect of tibial tarsal bone (Fig. 3.45). It is the largest bone of the hock joint. It is elongated, flattened from side to side; and consists of a body and medial process the sustentaculum tali. The body is enlarged at the proximal end to form the calcaneal tuber (The point of the hock) which gives insertion to the tendon of the muscles forming *achilies tendon*. Tendon of *gastrocnemius* muscle attaches on the caudal part, while tendon of *superficial digital flexor, biceps femoris* and *semitendinosus* muscles are attaches on anterior part of calcaneal tuber. Caudally tuber calcis is marked by a wide shallow groove which is coated by a cartilage in the fresh stale. The distal part of the body is compressed laterally, and bears a projection infront which articulates with the lateral malleolus, and the lateral surface of the tibial tarsal bone. The distal extremity bears a concave facet for articulation with the combined central and fourth tarsal bones. Medial surface of the body has a strong process, the sustentaculum tali, on its lower part which projects inwards. The plantar surface forms with the smooth medial surface of the body a groove for the passage of tendon of *deep digital flexor muscle*.

Central and fourth fused tarsal: - It is quadrilateral platelike bone extends throughout the entire thickness of the joint and articulates with all the other bones of the hock joint (Fig. 3.44). Dorsal surface presents two concave cavities separated by a ridge for articulation with condyles of tibial tarsal and has a spine at caudo-medial angle. Distal surface is at two different levels for fusion with fourth tarsal. Its medial side is on a higher level than the lateral and bears a large facet for articulation with the fused second and third tarsal bone and small facet behind for first tarsal bone. The lateral part bears a facet for articulation with the large metatarsal bone. Dorsal, lateral and medial surfaces are rough and continuous. Laterally there is a narrow undulating surface for articulation with the calcaneus bone. Caudal surface is irregular.

First tarsal:- It is small nodule-like bone and presents three articular surfaces. Proximal surface articulates with fused central and fourth tarsal. Dorsal surface articulates with fused second and third tarsal. Distal surface articulates with metatarsal bone.

Second and third fused tarsal:- It is a rhomboid piece of bone. It presents six surfaces. Dorsal and medial surfaces are nonarticular and continuous. Proximal, ventral, lateral and posterior surfaces are articular. The proximal surface is concavo-convex, and articulates with the central component. The distal surface is undulating and rests on the metatarsus. The fourth tarsal component articulates laterally with a small fact in front and first tarsal bone behind with small facet.

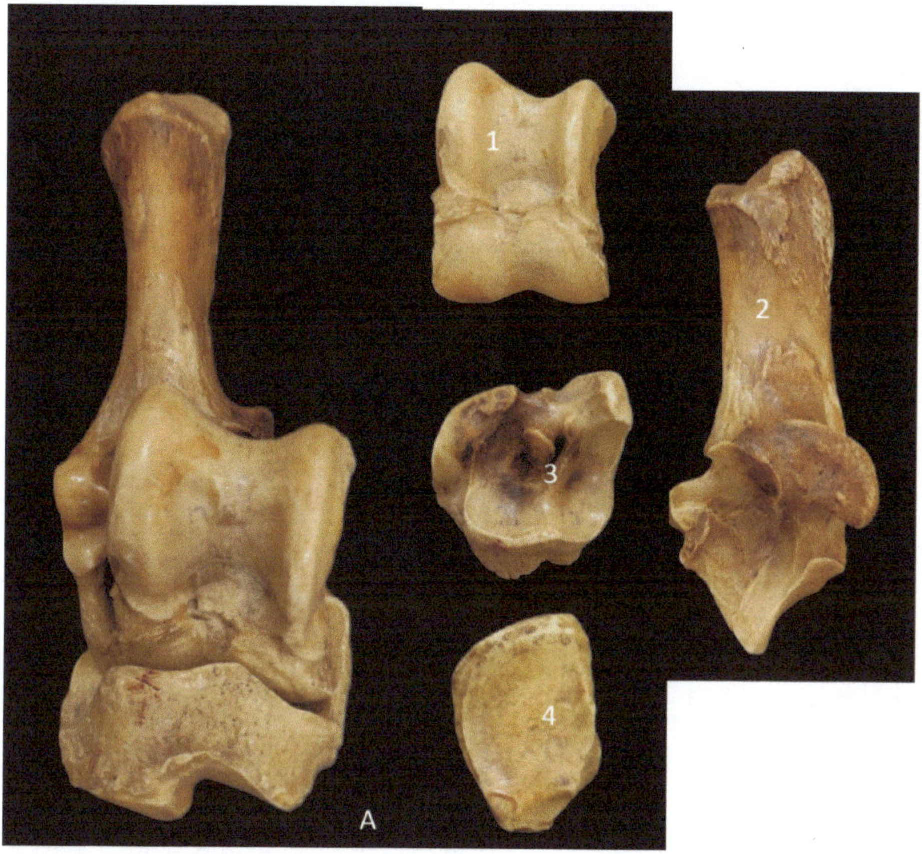

Fig. 3.44: Photograph showing the tarsus of Ox

1. Talus
2. Calcaneus
3. Central and 4th fused
4. 2nd and 3rd fused

Table 3.15: Comparative Table on Tarsus of Ox, Horse, Dog and Pig

Characters	Ox	Horse	Dog	Pig
Talus (Tibial tarsal) 1. Shape	Elongated pulley shaped bone, Larger and thinner than horse and flattened from before backward	Smaller and thicker than ox, not flattened from before backward	Consists of a body, neck and head	Axis is slightly oblique than ox
2. Proximal and dorsal surface	Continuous to form trochlea, The groove and ridges of the trochlea are straight. Articulates with cochlea of tibia and medial and lateral malleolus	Same as in ox but the groove and ridges of the trochlea are large and oblique.	Same as in ox but ridges of the trochlea are straight and articulates with tibia and fibula.	ridges of the trochlea are broad and sagittal
3. Distal surface	Shallow trochlea having two condyles which articulate with central and 4th fused tarsal	Condyles are absent, convex from, before backward, articulates mostly with the central tarsal by a large articular facet and 4th tarsal by small oblique, facet.	Form head which is directed little inwards and articulates with the central tarsal bone	Broad and rounded, articulate with central and 4th tarsal
4. Lateral surface	Irregularly concave and caudolaterally articulates with calcaneus with help of two facets	Marked by wide rough fossa, do not articulate with calcaneus,	Slightly concave, articulates with calcaneus	An articular facet for calcaneus
5. Medial surface	Bears a tuberosity at its upper part and flattened below	Contain two tuberosities, distal one is larger	In contact with medial malleolus	In contact with medial malleolus
6. Plantar surface	Large oval facet for articulation with calcaneus	Oblique, irregular and has four facets for articulation with the calcaneus	Two/ three facets for articulation with the fibular tarsal bone	Same as in ox
Calcaneus (Fibular tarsal)	Elongated short bone situated lateral and caudal aspect of talus, longer than horse	Not so long as in the ox and is stronger than ox	Same as in ox	Same as in ox

Characters	Ox	Horse	Dog	Pig
1. Body	Distal part of the body is compressed laterally, and articulates with the lateral malleolus and the lateral surface of the tibial tarsal bone.	Distal part of the body is not compressed laterally, and does not articulate with the lateral malleolus.	Dorso-medially articulates with talus and dorso laterally with fibula	Articulates with fibula, tibial and 4th tarsal, does not articulate with lateral malleolus
2. Lateral surface	Flattened above and has a rough prominence below	Same as in ox	Same as in ox	Same as in ox
3. Dorsal border	Smooth and rounded above, concave in its length	Same as in ox	Same as in ox	Same as in ox
4. Plantar border	Rough, Stright and broad, wider at the ends	Same as in ox	Same as in ox	Same as in ox
5. Sustenta-culum tali	A Prominent medial projection	More stronger and larger than ox , bears a facet infront for talus	Short/ indistinct	Comparatively Larger than ox and horse
6. Calcaneal tuber	A groove is present on the plantar part of calcaneal tuber.	Massive than ox, the groove on the plantar surface of calcaneal tuber is absent.	has a broad sagittal groove plantarly	deeply grooved plantarly
7. Processeus cochlearis	Less prominent than horse	Very prominent and bears facets on its medial and distal surfaces for articulation with the Talus bone.	Same as in ox	Same as ox
8. Distal extremity	Bears for articulation with the combined central and fourth tarsal bones	A concave facet Articulates with central tarsal bone	Articulates with the central and fourth tarsal bones	Same as in ox
Central tarsal	It is fused with 4th tarsal to form central and 4th fused tarsal bone.	It is a separate piece of bone which is irregularly quadrilateral and situated between the talus and third tarsal bone.	Not fused with 4th tarsal Separate bone	Not fused with 4th tarsal separate bone, thick and narrow transversely

Characters	Ox	Horse	Dog	Pig
	Quadrilateral plate like bone extends throughout the entire thickness of the joint and articulates with all the other bones of the hock joint.	Flattened from above downward situated between the talus and third tarsal bone.		
1. Proximal surface	presents two concavities separated by a ridge for condyles of tibial tarsal and has a spine at caudo-medial angle., laterally an undulating surface for articulation with distal end of calcaneus.	Concave from before backward and has a large facet for articulation with tibial tarsal, a non articular depression on is lateral part, lateral undulation is not present	Concave adopted to the head of the talus	Deeply concave and articulates with talus, plantar bears a large tubercle
2. Distal surface	Is has two different levels Medially it is higher level than the lateral and bears and large facet for articulation with the fused second and third tarsal bone and small facet behind for first tarsal bone. Laterally bears a facet for articulation with the large metatarsal bone.	Convex and is crossed by a non-articular groove which separated facets for articulation with the third and combined 1st and 2nd tarsal bones.	Articulates with the first, second, and third tarsals.	Convex and articulates with 3rd tarsal and fused 1st and 2nd tarsal
3. Plantar surface	Bears two tuberosities, lateral one is rounded and medial one is narrow and large	Bears two prominences separated by a notch.	bears two plantar tubercles.	Rough and non -articular

Characters	Ox	Horse	Dog	Pig
4. Medial surface	Dorsal, lateral and medial surfaces are rough and continuous.	Continue with dorsal surface, are convex, rough and non articular	Dorsal and medial surfaces are rough and continuous.	Rough and non -articular
5. Lateral surface		Oblique and bears the anterior and posterior facets for articulation with the 4th tarsal.	Articulates with 4th tarsal	Articulates with 4th tarsal
4th tarsal Shape		It is the lateral bone of the distal row and is equal in height to the central and third tarsal together. cuboid in shape and presents six surfaces	It is remarkably high, resembles a quadrangular prism,	Large,
1. Proximal surface		Convex from side to side and articulates chiefly with the fibular tarsal but to a small extent with tibial tarsal also.	Articulates with the calcaneus only	Articulates with talus and calcaneus
2. Distal surface		Rests on the large and small metatarsal bones.	Articulate with the fourth and fifth metatarsal,	Same as in dog

Characters	Ox	Horse	Dog	Pig
3. Medial surface		bears four facets for articulation with the central and third tarsal bones. It is crossed from before backward by a smooth groove, which by apposition with the adjacent bones forms the canal of the tarsus for the passage of the perforating tarsal vessels.	Articulates with central and third tarsal bones.	Same as in dog
4. Lateral surface		Dorsal, lateral and plantar surfaces are continuous and rough. A tuberosity behind gives attachment to the plantar ligament.	A groove for the tendon of the peroneus longus crosses its lateral and plantar surface, and above it there are one or two tubercles.	Same as in dog
1st tarsal 1. Shape	Small nodule like bone and presents three articular surfaces	Fused with 2nd tarsal to form 1st and 2nd fused tarsal bone Smallest and most irregular tarsal. very irregular shaped bone situated in the medio-plantar part of the distal row,	Flattened and irregularly quadrangular	High and narrow
2. Proximal surface	articulate with fused central and fourth tarsal bone	Articulate with Central tarsal bone by a concave facet	Articulates with the central tarsal bone	Articulates with central and 2nd tarsal bone
Dorsal surface	articulates with fused second and third tarsal	Articulates with third tarsal bone.	Non articular	Non articular
Distal surface	Articulates with metatarsal bone.	Distal surface has facets for large and small metatarsal	Articulates With the first metatarsal.	2nd metatarsal bone

Characters	Ox	Horse	Dog	Pig
Medial surface	Non articular	Medial surface has sharp notch indicating division of 1st and 2nd tarsal	Non articular	Non articular
2nd tarsal	Fused with 3rd tarsal to form 2nd and 3rd fused tarsal bone Rhomboid shaped, anterior and medial surfaces are nonarticular and continuous.		Smallest and is wedge shaped, it articulates proximally with central tarsal, laterally with 3rd tarsal and distally with the second metatarsal bone.	Small and prismatic, articulator with central tarsal proximally, 3rd tarsal dorsally and 1st tarsal plantarly, 2nd and 3rd metatarsal distally.
3rd tarsal 1. Shape		It resembles the central but is smaller and more triangular in outline. Situated between central above and large metatarsal below	It is also wedge shaped, the base being infront,	Larger, compressed proximal to distal. Wider dorsally and narrow plantarly.
1. Proximal surface	concavo-convex, and articulates with the central component.	Concave, crossed by non articular depression which divide this into two unequal facets for articulation with the central tarsal	Articulator with central tarsal	Articulator with central tarsal
2. Distal surface	Undulating and rests on the metatarsus.	Slightly convex and rests on the large metatarsal bone.	Articulates with the third metatarsal distally.	Articulates with 3rd metatarsal
3. Medial surface	Non articular	Deeply notched, has a single facet for articulation with 1st and 2nd fused tarsal.	Articulates with 2nd tarsal	2nd tarsal
4. Lateral surface	The fourth tarsal and first tarsal bone behind with small facet.	The lateral border is also divided by a notch into two parts and bears two diagonally opposite facets for articulation with the 4th tarsal.	4th tarsal laterally.	4th tarsal laterally.

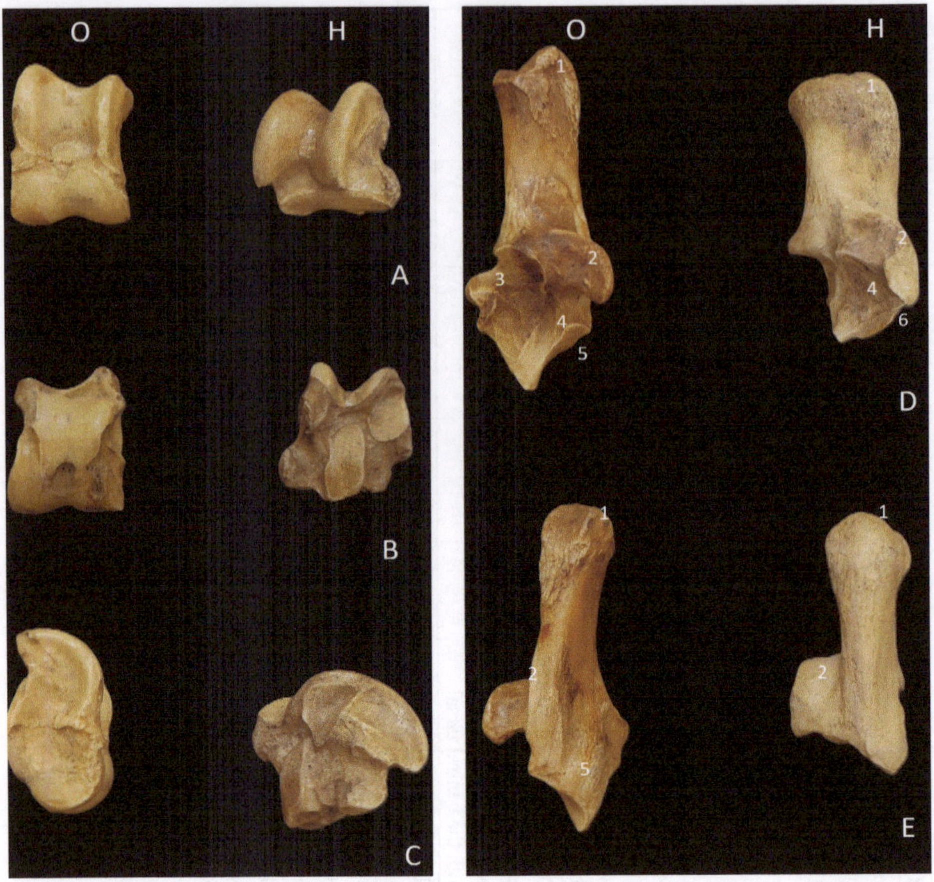

Fig. 3.45: Photograph showing Talus and Calcanius of Ox (0) and Horse (H) A. Dorsal view of talus, B. Plantar view of talus, C. Lateral view of talus , D. Medial view of calcaneus, E. Plantaro-lateral view of calcaneus

1. Calcaneal tuber
2. Sustentaculaum talli
3. Facet for lateral malleolus
4. Facet for talus
5. Facet for central and 4th fused carpal
6. Facet for central tarsal

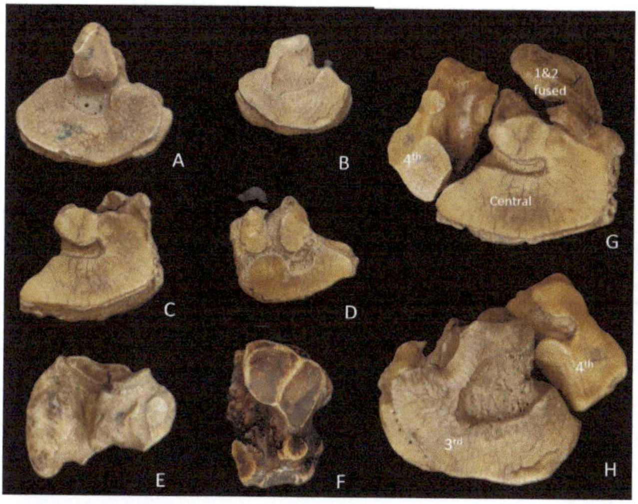

Fig. 3.46: Photograph showing the Tarsal of horse A. 3rd tarsal (proximal surface), B. 3rd tarsal (distal surface),C. Central tarsal (proximal surface), D. Central tarsal (distal surface), E. 1st & 2nd fused tarsal, F, 4th tarsal, G. Proximal view of tarsal of middle and row, H. Distal view of tarsal of distal row

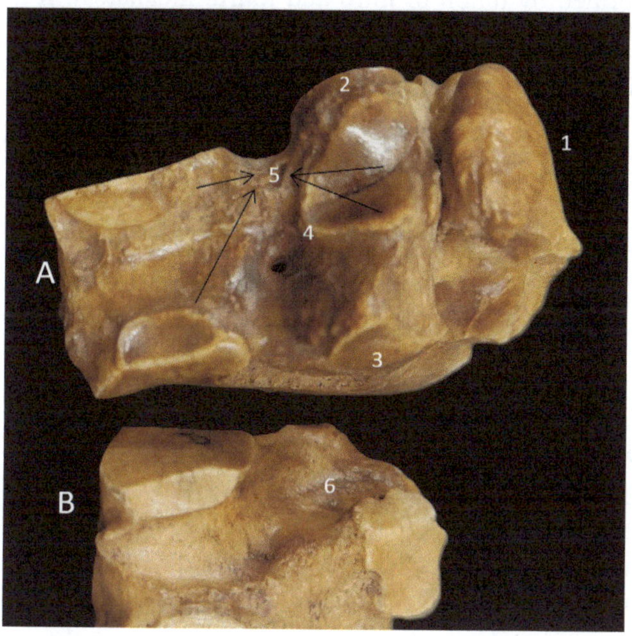

Fig. 3.47: Photograph showing lateral surface of central and 3rd tarsal bone (A) and medial surface of 4th tarsal bone(B) of horse

1. 2nd tarsal bone
2. Central tarsal
3. 3rd tarsal
4. Lateral surface of central and 3rd tarsal bones
5. Facets for 4th tarsal bone
6. 4th tarsal bone (medial view)

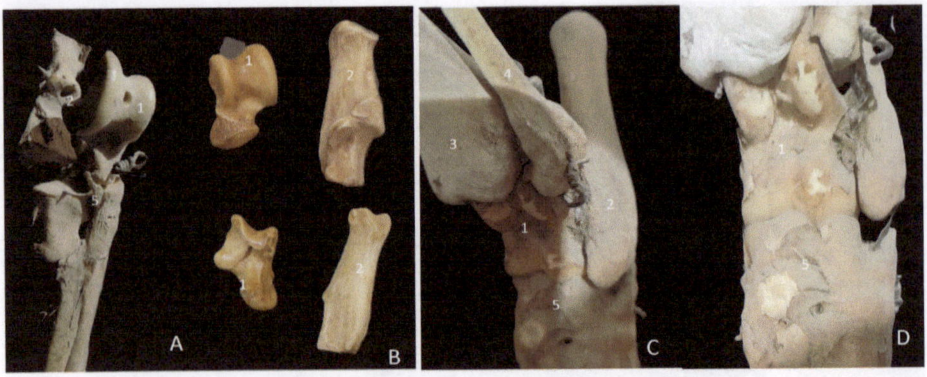

Fig. 3.48: Photograph showing A. tarsus of Dog, B. Dorsal view (upper) and plantar view (lower) of talus and calcaneus of Dog, C and D. Tarsus of Pig 1. talus, 2. calcaneus, 3. tibia, 4. fibula, 5. tarsal of distal row

e. Comparative Study of Metatarsal

Metatarsal of Ox

Large Metatarsal:- It is formed by union of third and fourth metatarsal. It is similar in structure as metacarpal except few points.

1. It is about one-seventh (1/ 7th) longer than the corresponding large metacarpal bone.
2. It is a long bone and directed slightly oblique downwards and a little forward directed.
3. Shaft is four sided and cylindrical in shape.
4. Dorsal longitudinal groove is more deep and wider than plantar groove.
5. The plantar surface present a shallow metatarsal groove.
6. The lateral and medial surfaces are smooth and concave from above downward.
7. The proximal extremity is quadriangular in outline and bears two large slightly concave facets for articulation with the bones of the lower row of tarsus.
8. The medial one articulates with the 2nd and 3rd fused and the lateral one with central and fourth fused tarsal bones.
9. There is a small facet behind the medial one for articulation with the 1st tarsal bone.
10. The medio-plantar angle of the proximal end bears a facet for articulation with the small metatarsal bone (2nd).

11. Distal extremity is similar to large metacarpal bone i. e. divided into two parts by sagittal notch.

Small metatarsal:- it is only one in number. It represents second metatarsal and situated medial to large metatarsal. It is rudimentary and quadrilateral in form. It articulates with large metatarsal by a facet at itsmedio-plantar aspect.

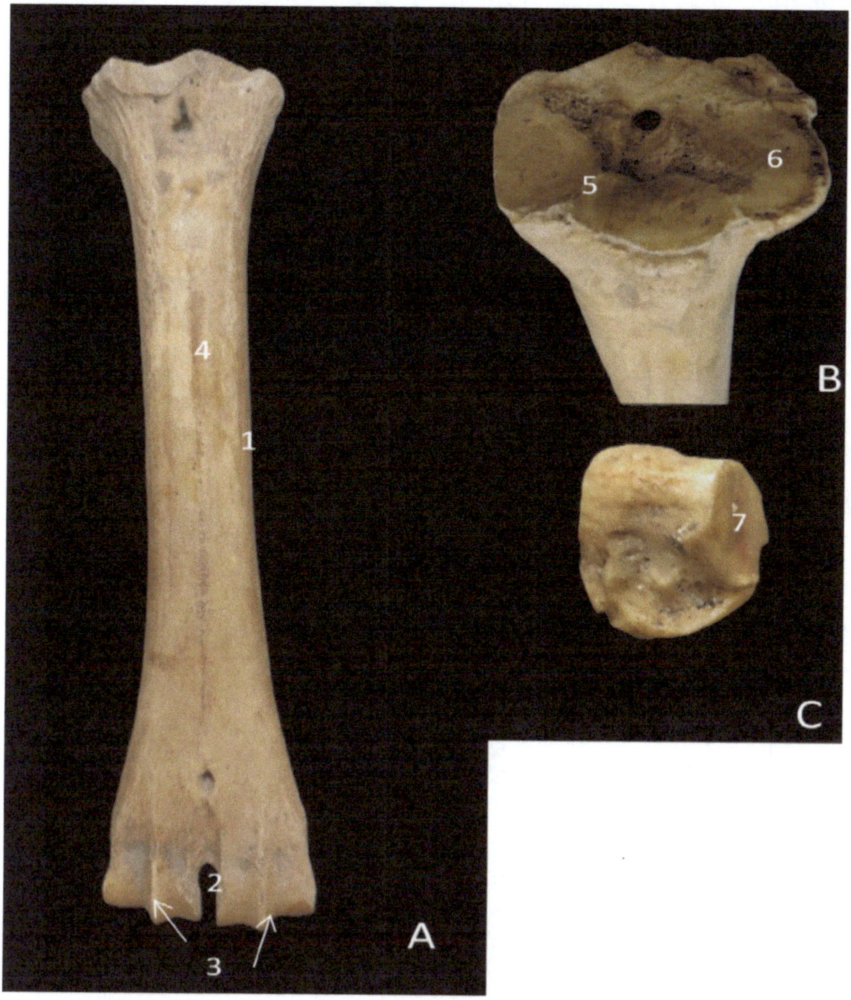

Fig. 3.49: Photograph showing the Left metatarsal of Ox A. Dorsal view, B. Proximal surface, C. Small metatarsal of ox

1. Dorsal surface
2. Sagital cleft
3. Condyles
4. Metatarsal groove
5. Facet for central and 4th fused tarsal
6. Facet for 2nd and 3rd fused tarsal
7. Facet for large metatarsal

Table 3.16: Comparative Table of Metatarsal of Ox, Horse, Dog and Pig

Characters	**Ox**	**Horse**	**Dog**	**Pig**
Number	3rd and 4th fused to form large metatarsal	3rd (only one large metatarsal)	5 metatarsals Similar to fore limb but thicker and longer	2 , 3, 4 and 5 which are Similar to fore limb but longer
Functional (carries a digit)	3 and 4	3	1,2,3,4,5	2,3,4,5
Non-functional	2nd	2,4 (Splint bone / small metatarsal)	None	None
No. of large metatarsal	One	One	Five	Two (3 and 4 metatarsals)
No of small metatarsal	One (5th metatarsal)	Two (2nd and 4th metatarsal)	0	2 and 5 (much smaller)
Direction	Slightly oblique, downward, and a little forward.	Same as ox	Same as ox	Same as ox
Surfaces	Compressed transversally and four sided	Body is more cylindrical and almost circular in cross section	1st is very small, blunt cone shaped 3 and 4 are four sided, rectangular 2 and 5 are three sided	3 and 4 are three sided, longer and three times thicker than 2nd and 5th metatarsal
Proximal extremity	Quadrangular in outline	Wider infront than behind	Elongated dorso-plantarly, have plantar projections 3rd and 4th have plantar facet for sesamoid bone	third and fourth metatarsal have plantar projections 3rd bear plantar facet for sesamoid bone
Dorsal longitudinal groove	Present	Not present	Not present	Not present

Characters	Ox	Horse	Dog	Pig
Articular surface and articulation	Proximally two large slightly concave facets - Medial facet articulates with the 2nd and 3rd fused and the lateral facet with central and fourth fused tarsal bones. A small facet behind the medial one for articulation with the 1st tarsal bone. Medio-plantar angle of the proximal end bears a facet for articulation with the small metatarsal bone (2nd).	Slightly concave, articulates with the third tarsal, but there is a lateral facet for the fourth, and usually a small facet caudo-medially for the second tarsal bone. caudally two pairs of facet for articulation with the small metatarsal bones	Corresponding tarsal bone	Articulates with each other and corresponding tarsal bone
Distal extremity	Divided into two parts by a sagittal cleft	undivided	Head like articular surface	Condyloid dorsally, trochlear palmarly
Small metatarsal	In the form of sesamoid bone	Little longer than small metacarpal	Absent	2 and 5 Same as small metacarpal
Medial small metatarsal (2nd)	rudimentary , about 2 cm in height and width, quadrilateral, articulates with large metatarsal by a facet at its anterior aspect	Slightly longer than 5th small metatarsal. It has two facets dorsally for large metatarsal, and two facets on upper face for the first and second fused tarsal		
lateral small metatarsal (5th)	Absent	Slightly smaller than 2nd metatarsal. Facets are similar to 2nd metatarsal.		

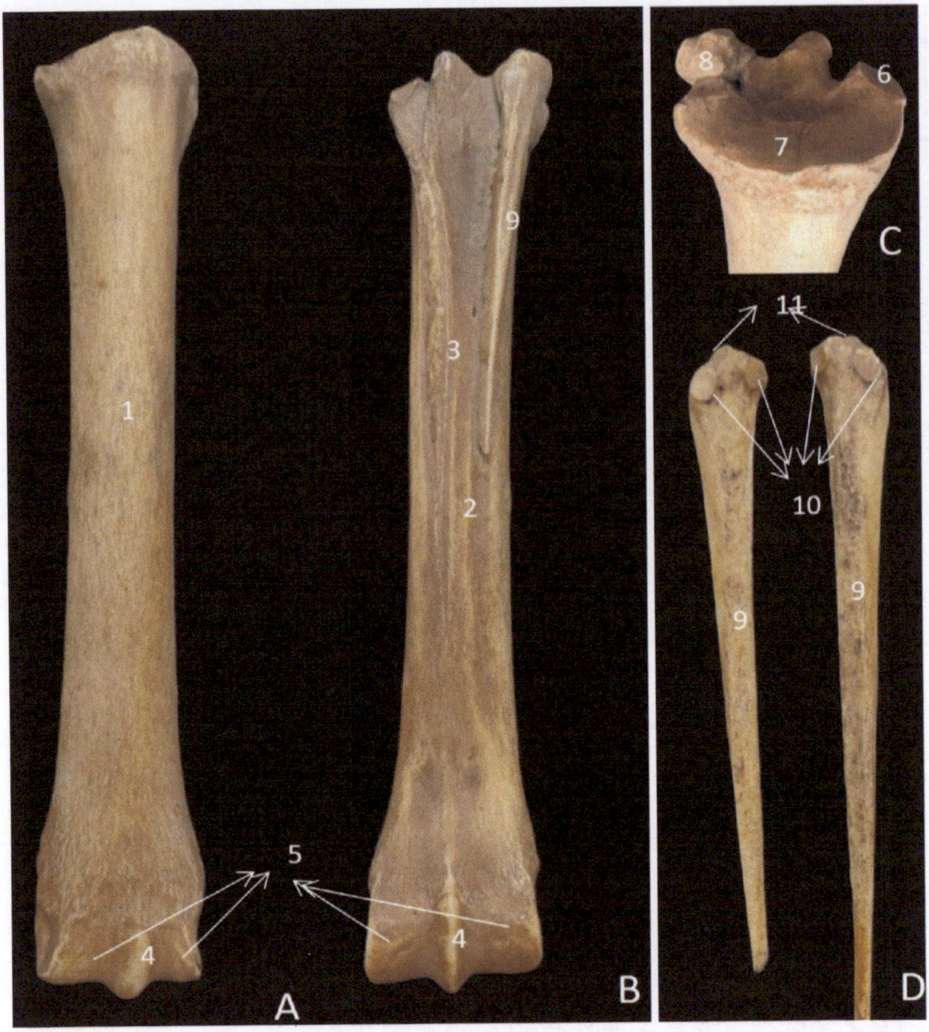

Fig. 3.50: Photograph showing the Right Metatarsal of Horse A. dorsal view, B. Plantar view, C. Proximal extremity, D. Splint (small metatarsal) bones

1. Dorsal surface
2. Plantar surface
3. Groove for splint bone
4. Sagital ridge
5. Condyles
6. Facet for 1st & 2nd fused tarsal
7. Facet for 3rd tarsal
8. Facet for 4th tarsal
9. Splint bones
10. Facet for large metatarsal
11. Facet for tarsal

f. Comparative Study of Digits, Phalanges and Sesamoids

Digits, Phalanges and Sesamoids of Ox

Digits: total four digits are present out of which 2nd and 5th digits are non-functional and forms dew claw. 3rd and 4th digits are functional digits which are well developed and formed by three phalanges each and sesamoid bones. The arrangement of phalanx and sesamoids are similar to fore limb.

Phalanges and sesamoids:-They are almost similar as described in forelimb. 1st and 2nd phalanxes are somewhat shorter than fore limb.

Digits, Phalanges and Sesamoid Bones of Horse

Similar to forelimb. Following differences are present between them:

1. The phalanges of hind limb form an angle of about 5° with ground whereas in forelimb no angle is formed between phalanges and ground.
2. In hind limb, the proximal or First Phalanx is a little shorter, wider above and narrower below than the 1st phalanx of forelimb.
3. The Middle or Second Phalanx is slightly longer and narrower than the 2nd phalanx of forelimb..
4. The distal or Third Phalanx is narrower than 3rd phalanx of forelimb. The angle of inclination of dorsal surface is a little greater (about 5 °), the plantar surface is more concaved, and the angles are less prominent and closer together.
5. The Proximal Sesamoid are a little smaller than the sesamoids of forelimb, except in thickness.
6. The Distal Sesamoid is narrower and shorter.

Digits, Phalanges and Sesamoid Bones of Dog

In dogs, the first digit is often absent. But, when present, its development varies and it contains one or two phalanges. In case of very large dogs, a sixth digit may be present which does not articulate with the metatarsus, but is attached by fibrous tissue. The phalanges of the other digits resemble those of the fore limb.

Digits, Phalanges and Sesamoid Bones of Pig

Similar to fore limb. The proximal and middle phalanges are little longer and narrower than those of the phalanges of fore limb.

4

The Skeletal of Fowl

In birds the skeletal system consists as similar to all higher vertebrates. In adult birds the medullary cavity of long bones and medullary space of spongy bones are poorly filled with blood forming bone marrow. It is filled with air as the diverticulum of pulmonary air sac system enter these cavities through pneumatic foramen during early stages of development and thus the bones become pneumatic bones. The bones of the skull are not connected with air sacs but few bones also contain air because they are ventilated through the mouth cavity. The thoracic limb of birds is modified for purposes of flight and is commonly known as the wing. Shoulder girdle is formed by scapula coracoid and clavicle.

Bones of Forelimb

Scapula

The scapula is an elongated sward shaped bone situated parallel to vertebral column and reaches almost to the cranio-lateral border of the ilium. It articulate cranially with coracoid and furculum by acromion and Clavicular process, respectively to form glenoid fossa for the head of humerus and a space present between them called foramen triosseum which gives passage to the tendon of supracoracoideus muscle (Fig. 4.1 and 4.2).

Coracoid

It is rod like strongest bone of shoulder girdle. Ventral extremity is flat and expanded with a saddle shape structure and articulate with the sternum at cylindrical articular protuberance, stero-coracoidea located at the cranial border of sternum. At this end there is also an opening leads into clavicular air sac. Dorsal end is prolonged hook like process, tuberositas furcularis into which medial surface of clavicle is fixed by connective tissue. Below this, there is depression which form glenoid cavity for articulation with head of humerus (Fig. 4.1 and 4.2) .

Clavicle

The clavicle is another rod like, thin and slightly bend bone. Upper end (dorsal) is expanded and connected with coracoid and scapula to form glenoid cavity. Ventrally two clavicles are united with an acute angle and form furcula which act as a spring between two shoulder joints and gives support to the wing. Cranially it is connected with the rostrum of the sternum with ligament.

Humerus

The humerus is stout and slightly curved pneumatic bone. It lies almost horizontal to the lateral wall of the thorax and reaches from shoulder joint to the cranial border of ilium in folded wing. Proximally head of the humerus articulates with scapula and coracoid. The head is oval in form. On each side of the proximal end there is a prominent tubercles lateral (dorso-laterally) and medial tubercles (ventro-medially). A large opening, Pneumatic foramen, is situated medially close to the medial tubercle below the head. It communicates with the cavity inside the bone and serving as an entrance for air from the Clavicular air sac. The distal extremity of the bone presents two convex articular areas, condyles (ulnar and radial) and large ulnar and small radial epicondyles (Fig. 4.1).

Radius And Ulna

The ulna is comparatively massive among the two bones of the forearm. Both the bones are in contact at two ends and enclose a wide interosseous space. The radius is rod shaped and located cranial to the ulna. The proximal end of the radius can be distinguished by a concave articular surface of rounded outline. The shaft of the ulna is cylindrical and curved. Ulna carries a small projection, the olecranon, at its proximal and larger end. Semilunar notch is well marked (Fig. 4.1).

Carpals

The carpus of the adult fowl contains two bones only, a radial and an ulnar carpal similar to the proximal row of mammalian carpal bones. The radial carpal bone articulate with radius and ulna both whereas, ulnar carpal articulates with ulna only. In the embryo the carpals of distal row distal row are represented by cartilaginous nodules; but later on during development they get fused with metacarpal to form carpo-metacarpal bone.

Metacarpal

There are three (I, II and III) metacarpals fused with distal row of carpal bones forming irregular D shaped structure called as Carpometacarpal bone. The second and third carpo-metacarpal bones are blended at their ends only, the intervening shafts being separated by a narrow, elongated interval. The first metacarpal bone is represented by a small, short projection at the proximal end of the second carpometacarpal bone.

Digits

There are 3 digits. First/ pollex and second have 2 phalanges and third has one phalanx. They extend in a radiating manner from the corresponding area of carpometacarpus. Sesamoids are absent.

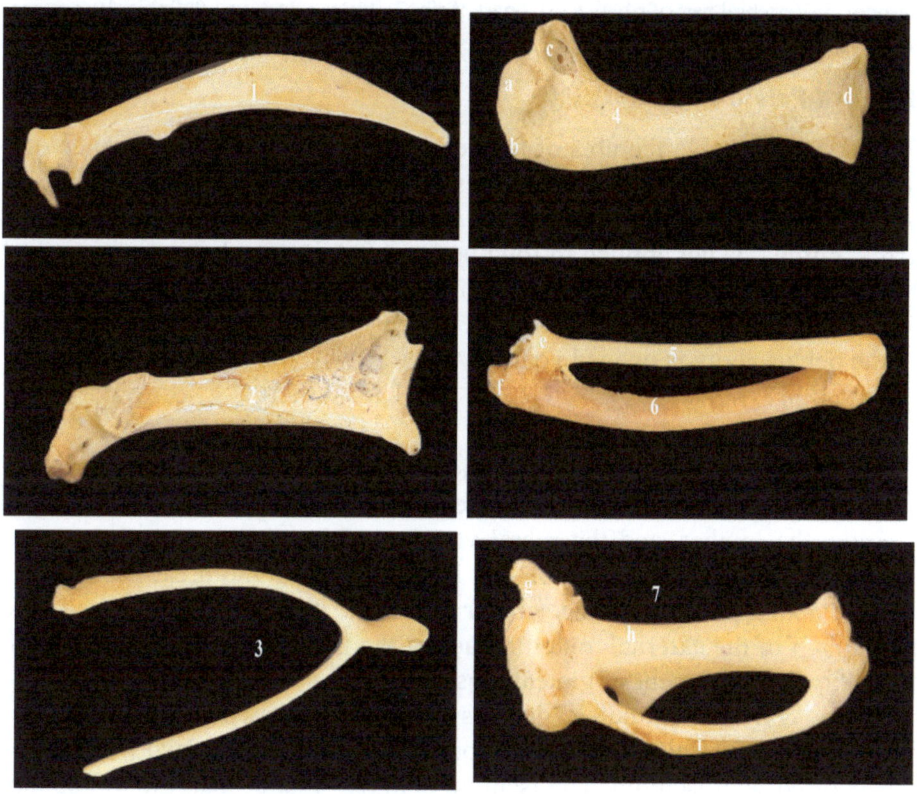

Fig. 4.1: Photograph showing the Wing bones of Fowl: 1 .Scapula, 2. Coracoid, 3. Clavicle, 4. Humerus, 5. Radius, 6. Ulna, 7. Carpo-metacarpal, a. Head of humerus, b. Tubercle, c. Pneumatic foramen, d. Condyles, e. Head of radius, f. Olecranon of ulna, g. I metacarpal, h. II metacarpal and i. Ill metacarpal

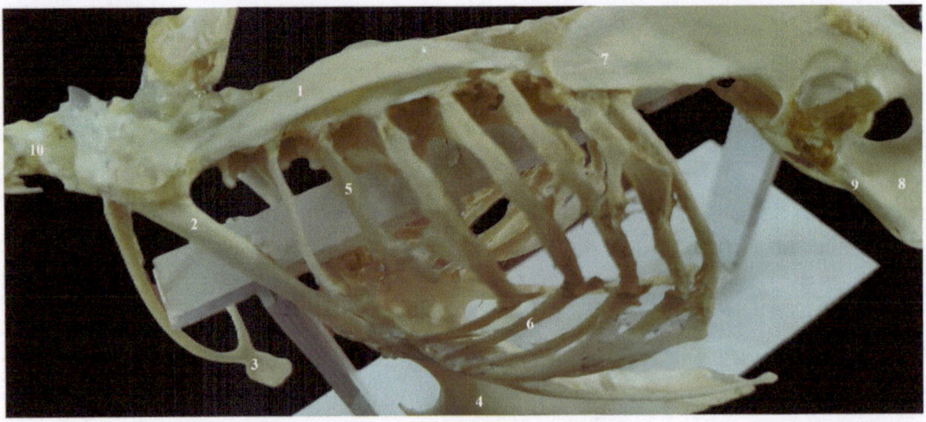

Fig. 4.2: Photograph showing different bones of fowl 1 .Scapula, 2. Coracoid, 3. Clavicle, 4. Sternum, 5. Vertebral part of rib, 6. Costal part of rib, 7. Ilium, 8. Ischium, 9. Pubis, 10. cervical vertebra

Bones of Pelvic Limb

The most remarkable features of the pelvic limb, or leg, are the absence of a ventral union between the two hip-bones, firm and extensive fixation of the hip-bone to the vertebral column, and the absence of an independent tarsus. The term " girdle " is not applied in fowl because Pelvic symphysis is absent and the bones of pelvis does not united ventrally, instead of the bones of opposite sides of the body meets and united with the synsacrum of the vertebral column. So the pelvic cavity is opened ventrally.

Os Coxae

The ilium is the largest segment, and is joined firmly to the transverse processes of synsacrum that is fused part of the vertebral column which contains the last thoracic, the lumbar, and the sacral vertebrae. Ilium is elongated extend to entire length of hip bone (Fig. 4.3). It is differentiated into preacetabular and post acetabular parts. the preacetabular part of the ilium is Shovel shaped. The cranial half of the external/ outer surface of the ilium is in the form of a deep concavity for lodgement of gluteal muscles, while the caudal half of the same surface is convex. The pelvic/ inner surface presents two concavities, renal fossa for kidney. Half of the lateral border is free in its cranial part caudal half is fused with ischium. The ischium is much smaller than the ilium, and, in the adult, continuous therewith except at a large, ovoid opening, the sciatic foramen. Sciatic foramen is formed between postacetabular part of the dorsal border of Ischium and ventral border of ilium. Pubis is thin, long and slender shaped bone running along the ventral border of ischium. Cranial end of pubis has muscular process. The free caudal end of the pubis projects backwards for a little distance beyond the Ischium. An oval obturator foramen separates the pubis and ischium close to the acetabulum. Acetabulum is relatively large and deep and perforated. In fowl it is formed by olium and Ischium bone only because pubis get fused with iascium below the cetabulum. Ilioischiatic foramen is the extra foramen between ilium and ischium. Immediately above the acetabulum a projection of the ilium carries an articular surface covered with cartilage, antitrochanter for contact with the trochanter major of the femur.

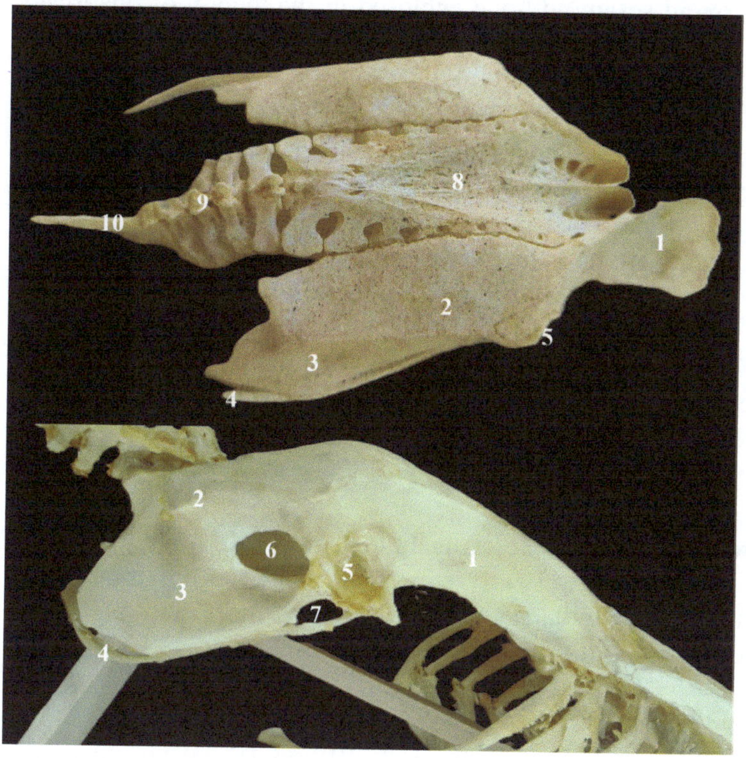

Fig.4.3: Photograph showing the oscoxae, synsacrum, and coccygeal vertebrae of fowl

1. Preacetabular part of ilium
2. Post acetabular part of ilium
3. Ischium
4. Pubis
5. Acetabulum
6. Ischiaticforamen
7. Obturator foramen
8. Synsacrum
9. Coccygeal vertebrae
10. Pygostyle

Femur

The femur is stout, cylindrical, and somewhat bent bone. In standing condition it is cranio-distally, laterally directed. It's proximal extremity comprises of head and greater trochanter. The head is comparatively small and rough.

Greater trochanter/ trochanter major is present lateral to the head in the form of irregular projection. Trochanter minor is in the form of small protuberance and situated medially below the neck. The distal end carries a deep, pulley-shaped surface for the patella, and two convex condyles for articulation with the bones of the leg. The lateral condyle presents, on its lateral aspect, a groove for the head of the fibula (Fig. 4.4).

Patella

It is thin, small, ovoid sesamoid bone enclosed in the tendon of quadriceps femoris muscle. Its width is comparatively more.

Tibia and Fibula / Shank Bone

Shank bone is consists of tibia and fibula. Tibia is longest bone in the body and much longer bone than the femur. Tibial crest is prominent. Proximal extremity is large and irregular. It presents small external and large internal condyles for femur. Distal end is condyloid. It has two articular elevations separated by a smooth groove. Proximal row of tarsal bones is fused with distal end of tibia forming a tibiotarsus. The fibula is thin rod-shaped long bone remains attached to the lateral part of the proximal end of tibia by a flattened head, which articulates with the lateral condyle of the femur. Distally, it is pointed and extends upto the distal third of tibia and fuses with it (Fig. 4.4).

Tarsal

In the adult there is no independent tarsus. In the embryo there are indications of two rows of elements; but in early embryonic stage the proximal row is fused with tibia forming tibiotarsus and distal row is fused with metatarsal forming tarsometatarsus.

Metatarsal/ Tarso-Metatarsal

It is single long bone representing fused II, III and IV metatarsal bones. Proximal end is irregular due to fusion of distal row of tarsal bones. It has concave facet for tibiotarsus. Distal end has three trochlea for 2nd, 3rd and 4th digits. In the male a conical and slightly curved projection, a bony spur arises from the medial side at level of distal third. A rudimentary first metatarsal bone is located at caudal side of the united bone and is connected by ligament. The fifth metarsal bone is represented in the embryo by a small cartilaginous nodule which soon disappears.

Digits

There are four digits. First arises from first metatarsal directed backward and inward and has two phalanges. Other three main digits are directed forward in a diverging manner and has one additional phalanx to their number. the second digit has three; the third has four ; and the fourth has five phalanges (Fig. 4.4).

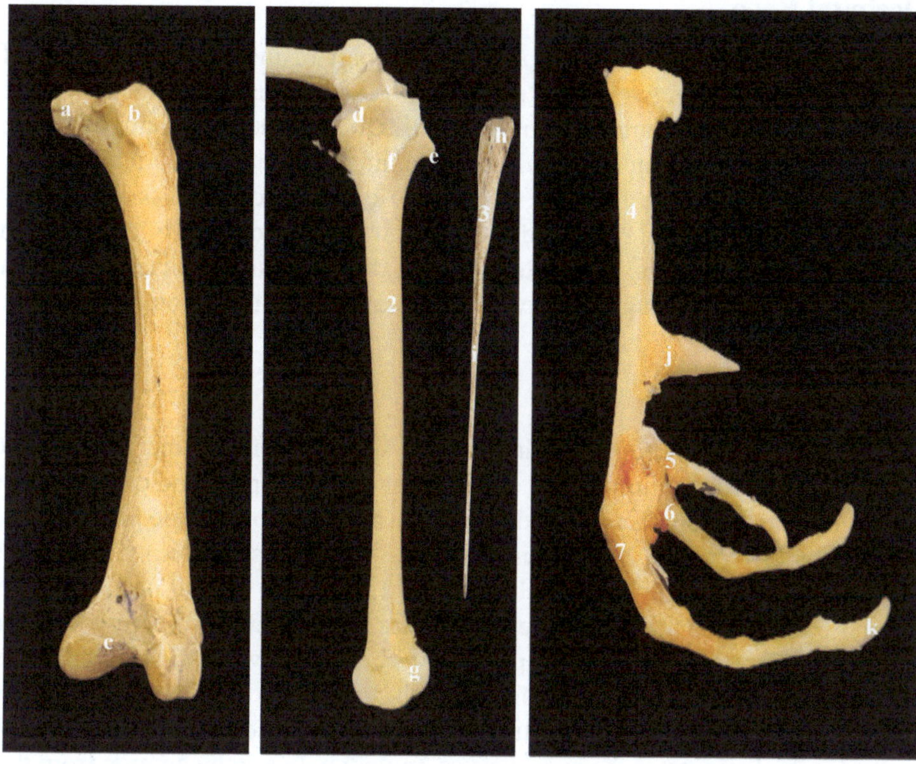

Fig. 4.4: Photograph showing the bones of pelvic Limb of fowl: 1. Femur, 2. Tibio-tarsus , 3. Fibula , 4. Tarso-metatarsal, 5.1st digit, 6. 2nd digit, 7. 3rd digit: a. Head of femur, b. Trochanter major, c. Condyles of femur, d. Lateral Condyle of tibia , e. Medial condyle of tibia, f. Tibial crest, g. Trochlea of tibio-tarsaus, h. Head of fibula, i. Body of fibula, j. Spur and k. Claw

The Skull

All the bones of the avian skull fuse during the embryonic period before hatching, as the sutures become ossified and the bones unite completely. The presence of large orbits, separated only by a thin septum formed by the sphenoid and ethmoid bones, allows a clear distinction between the rounded cranium and the pointed, conical facial region. The cranial portion of the skull encloses the brain and houses the organs of hearing, while the facial portion, situated anterior to the cranium and movably articulated with it, includes the skeletal framework of the jaws and the hyoid apparatus. A prominent feature of the skull is the presence of a large, hemispherical tympanic cavity on each side of the posterior part of the cranium, which forms the drum of the ear when the soft tissues are in place.

Occipital Bone

The occipital bone forms the most posterior part of the skull. It consists of four Parts viz. basilar, two lateral masses and a squamous part. Together, these components enclose a large opening known as the foramen magnum, by which the cavity of the cranium is placed in communication with the vertebral canal. Immediately below the foramen is a single occipital condyle is formed by basioccipital and lateral parts and articulate with deep concavity of atlas and odontoid process of axis. Each lateral part of the occipital bone is pierced by a common opening leading into the jugular foramen and the carotid canal (Fig. 4.5).

Sphenoid Bone

The sphenoid bone is roughly triangular in shape and forms the greater part of the base of the cranium. It lies cranial to the basioccipital and caudal to the ethmoid bone. It consists of a caudal portion (basisphenoid) made up of a body and a pair of temporal wings, and a cranial part (Presphenoid) represented by the body alone. The body of the sphenoid forms the central portion of the cranial base. On each side of the body there is a slightly projecting smooth surface for articulation with the pterygoid bone. The two optic nerves leave the cranium through a single opening, which represents the two optic foramina of mammals. This common aperture lies opposite the edge of the interorbital septum, which separates the optic nerves immediately as they emerge from the cranium. The temporal wings extend laterally from the body of the sphenoid and contribute to the formation of the cranial floor and lateral wall. They provide areas of articulation for surrounding bones, including the temporal and pterygoid (Fig. 4.5).

Parietal Bone

The parietal bones in the fowl are paired, broad, and short elements that form part of the roof and lateral walls of the cranium. They are situated between the frontal bones cranially and the occipital bone caudally. Each parietal bone articulates cranially with the frontal bone, caudally with the squamous part of the occipital bone and laterally with the squamous temporal bone. The external surface of the parietal bone is smooth and convex, contributing to the rounded contour of the cranium, while the internal surface is concave, forming part of the cranial cavity that houses the brain.

In the fowl, the interparietal bone (commonly present as a separate median element in mammals) is absent. The two parietal bones may fuse in the midline during later development, forming a continuous cranial roof.

Frontal Bone

The frontal bone is the largest bone of the cranium in the fowl. It forms a major portion of the roof of the cranial cavity, as well as parts of the roof of the orbits and the upper surface of the face. It lies between the parietal bones caudally and the nasal bones rostrally. The frontal bone can be divided into three parts viz. Frontal part, Nasal part, and Orbital part. The Frontal Part forms the main cranial roof. It articulates caudally with the parietal bones and laterally with the squamous temporal and zygomatic process of the malar bone. Its internal surface is smooth and concave, forming part of the cranial cavity that lodges the forebrain. The nasal part extends forward between the nasal bones. It contributes to the formation of the upper beak region and articulates with the nasal and premaxillary bones cranially. The orbital Part forms the roof of the orbit. It bears a strong supraorbital process which provides attachment for orbital structures and protection for the eye. The lateral border of the frontal bone joins the zygomatic process of the malar and the squamous part of the temporal bone, completing the upper rim of the orbit (Fig. 4.5).

Temporal Bone

The temporal bone of the fowl forms part of the lateral wall and base of the cranium and is closely connected with the basioccipital and sphenoid bones. It contributes to the housing of the organ of hearing and to the articulation of the quadrate bone, which connects the skull with the mandible. It is composed of two main parts viz. Squamous temporal, and Petrous temporal. The squamous portion lies on the lateral aspect of the cranium. It bears two distinct processes, the orbital and the zygomatic processes. The zygomatic process extends forward to meet the frontal bone, while the supraorbital process fuses with the corresponding process of the frontal bone. The fused supraorbital process forms part of the articular facet for the quadrate bone, which permits cranial kinesis (movement of the upper beak). The petrous temporal portion is concealed by adjacent bones. It forms the floor of the tympanic cavity and contributes to the articular facet for the quadrate bone.

Ethmoid Bone

The ethmoid bone of the fowl is a small, median bone situated between the orbits, forming part of the floor of the cranial cavity and the roof of the nasal cavity. It lies cranial to the sphenoid and caudal to the nasal bones.It can be divided into two main parts horizontal plate and perpendicular plate. The horizontal plate corresponds to the cribriform plate of mammals. It is perforated on each side by several minute foramina that allow the passage of olfactory nerves from the olfactory bulbs of the brain into the nasal cavity. The perpendicular plate forms the interorbital septum, which separates the two orbits. Within this septum, there is an opening that communicates with the single optic foramen, through which the optic nerves pass. In fowls, the lateral masses and the complete horizontal part seen in mammals are absent or poorly developed.

Facial Bones

Maxilla Bone

The maxillae are paired, small, rod-shaped or plate-like bones articulate cranially with the premaxilla and nasal bones, and caudally with the malar (jugal) and palatine bones. Palatine processes do not meet in the middle line **but** join the palatine and vomer posteriorly (Fig. 4.5).

Premaxilla Bone

It forms the base for horny beak. It has a body and three pairs of processes of frontal, maxillary and palatine bone. the *palatine process* which meets the palatine, the *maxillary processes* which extend backward and meet the maxilla and the *frontal processes* which extends dorsally and backward along the median line and between the nasal bones and meet the frontals.

Nasal Bone

The nasal bones are thin, elongated plates forming the dorsal and lateral boundary of the nasal chambers. It has a body and three processes viz. *frontal, premaxillary* and *maxillary.* These circumscribe the anterior nares with the premaxilla and the maxilla. They are notched at the cranial end by the opening into the nasal chamber. The nasal bones are loosely connected to the frontal bones, permitting movement during cranial kinesis (Fig. 4.5).

Palatine Bone

The palatine bones are paired, elongated bones that form the caudal part of the hard palate and the lateral walls of the nasal cavity. They are wide caudally and narrow cranially, meeting in the midline to form a median suture. Each palatine articulates caudally with the sphenoid and pterygoid bones, and cranially with the maxilla and premaxilla. They also help to form the posterior nasal openings (choanae) and the roof of the mouth (Fig. 4.5).

Lacrimal Bone

It is triangular plate of bone situated at the cranial margin of the orbit. Its medial edge meets the frontal. The apex, pointed downward, is continued by a long curved process. It helps form part of the cranial orbital wall and contributes to the passage between the orbit and nasal cavity (Fig. 4.5).

Malar or Jugal Bone

A very thin and elongated slender bone is composed of a cranial jugal bone and a caudal quadratojugal. It forms a connecting link between the face and the cranium on each side of the skull and articulates caudally with the quadrate and cranially with the maxilla (Fig. 4.5).

Pterygoid Bone

The pterygoids are slender, rod-shaped bones that extend between the rostrum of the sphenoid medially and the quadrate bone laterally. They form an important link between the palate and the cranium, transmitting movement from the quadrate to the upper jaw during beak motion.

Turbinate Bone

These are three cartilaginous ridges throughout the life *viz. anterior, middle* and *posterior*. The middle is the largest and posterior is the smallest.

Vomer Bone

The vomer is a slender, partly cartilaginous rudimentary bone that extends forward from the rostrum of the sphenoid and is attached above to the septal cartilage in its cranial part to form a part of the nasal septum.

Quadrate Bone

It is irregularly quadrangular and intervenes between a concavity on the temporal bone, just in front of the tympanic cavity, and the mandible. It has three processes viz. otic, articulatr process and orbital process. The otic process forms a movable joint with the squamous temporal bone. The orbital process pointed towards the orbit and acts as muscular lever. The strong articular process has a condyle which articulates with the mandible. The presence of the quadrate bones permits of the mechanical elevation of the skeleton of the upper part of the beak whenever the mandible is depressed (Fig. 4.5).

Mandible

It is single bone resulted from rostral fusion of two halves right and left . It presents five elements viz. dentary, splenial, supraangular, articular and angular. The articular process forms the caudal part of the mandible and is expanded. It carries a slightly concave surface for articulation with the quadrate bone, behind which the lower border of the bone is continued as angular process which curves in an upward direction. It has an internal and posterior part. The posterior third or more of the upper border of the mandible is formed by the supra-angular, on which there is a very small coronoid process a little anterior to the articular surface for the quadrate. The splenial is a thin plate of bone lying along the inner face of the mandible. The largest element of the mandible is the dentary, which forms the anterior half of the jaw and fuses firmly with its fellow of the opposite side (Fig. 4.6).

Hyoid

It consists of an unpaired rodlike body, basihyoid and two rami. The body prolongs both rostrally and caudally.

At the rostral end of body an articular surface is present for entoglossal bone, is sagittate, double in fowl and embedded in tongue. The caudal part of the body is connected with urohyoid by a cartilage. Urohyoid is partly bony and partly cartilaginous. The lateral part of the body caries articular surface for ramus. The ramus is in the form of bony rod which forms movable joint with the body.

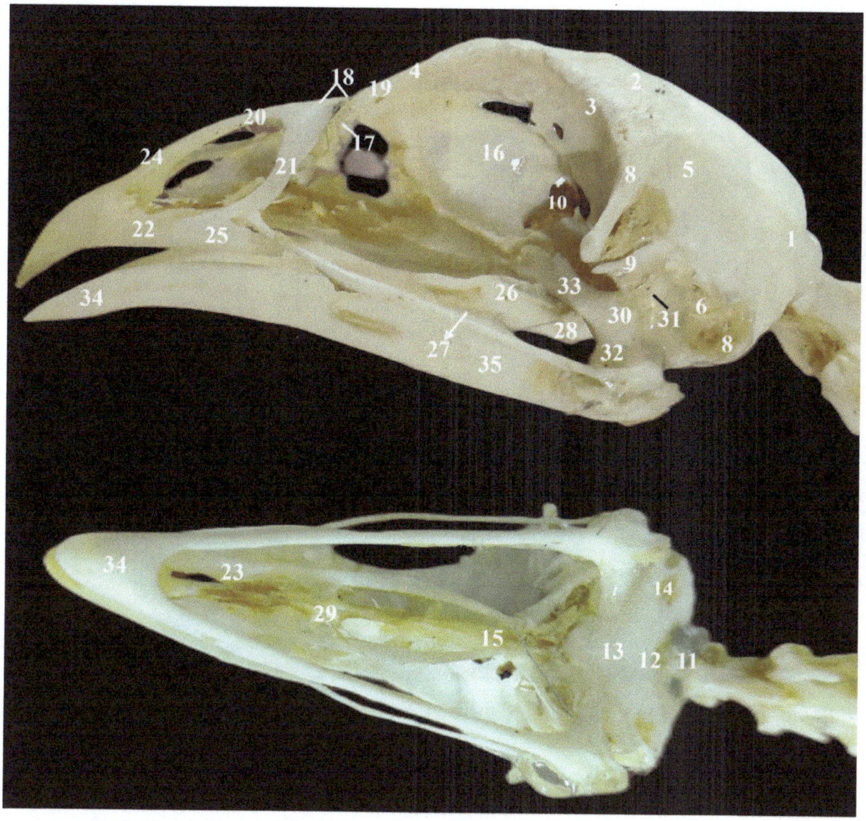

Fig. 4.5: Photograph Showing the Skeleton of Head of Fowl

1. Parietal bone
2. Frontal part of frontal bone
3. Orbital part of frontal bone
4. Nasal part of frontal bone
5. Squamous part of temporal bone
6. Petrus part of temporal bone
7. External acoustic meatus
8. Orbital process of temporal bone
9. Zygomatic process of temporal bone
10. Optic foramen
1 1. Foramen magnum
12. Occipital condyle
13. Basi sphenoid
14. Jugular foramen
15. Presphenoid
16. Interorbital septum
17. Lacrimal bone
18. Nasal bone
19. Frontal process of nasal bone
20. Premaxillary process of nasal bone
21. Maxillary process of nasal bone
22. Maxillary process of premaxilla
23. Palatine process of premaxilla
24. Frontal process of premaxilla
25. Maxilla bone
26. Palatine bone
27. Mallar bone
28. Pterygoid bone
29. Vomer bone
30. Quadrate bone
31. Otic process of quadrate bone
32 Articular process of quadrate bone
33. Orbital process of quadrate bone
34. Mandible
35. Supra angular bone

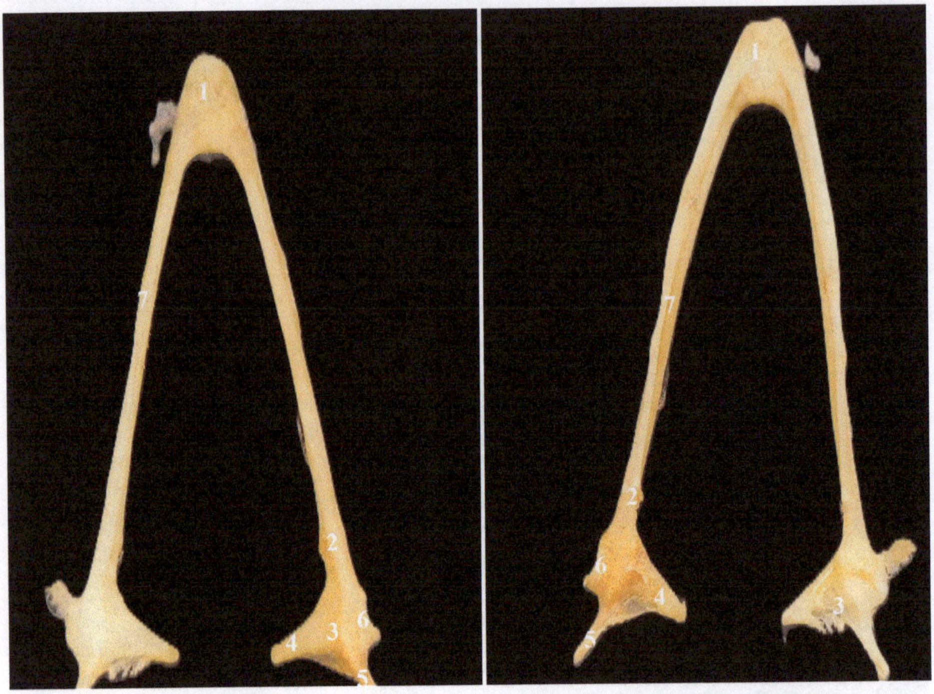

Fig. 4.6: Photograph showing the mandible of fowl A. Ventral view, B. Dorsal view

1. Dentine
2. Supra angular part
3. Angular part
4. Internal part of angular process
5. Posterior part of angular process
6. Articular process
7. Splenial

The Vertebral Column

The vertebral column may be divided into cervical, thoracic, lumbar, sacral, and coccygeal regions; but the limits of these are not as well defined as in mammals.The vertebral column, or back-bone, of the fowl is remarkable for several reasons. The part of vertebral column which forms the skeleton of the neck i. e. cervical vertebra is long and very freely movable; whereas the rest of the vertebra viz. thoracic, lumbosacral and coccygeal are rigid owing to the extensive fusion which takes place between the bones composing it. Some of the thoracic vertebrae are firmly united to form a firm foundation may for the wing. Most extensive fusion occurs in the lumbar and sacral regions. The greater part of hip bone is immovably fixed to the vertebral column by a bony union through out its length. Each vertebra consists of a rod-like body, surmounted by a vertebral arch ; these circumscribing a vertebral foramen which forms part of the vertebral canal of the articulated skeleton. The ends of the body of a vertebra have saddle-shaped surfaces for articulation with adjacent vertebrae. The cranial surface is concave from side to side and convex in a dorso-ventral direction. The caudal surface is convex from side to side and concave in dorso-ventral direction. In many vertebrae the ventral aspect of the body carries a median, flattened process. In a typical vertebra the arch is produced dorsally into a spinous process, and carries two cranial and two caudal articular processes with smooth surfaces for contact with similar processes on neighbouring bones. From each side of a vertebra projects a transverse process attaining a considerable length in the thoracic, lumbar, and sacral regions.

Cervical Vertebrae

The cervical vertebrae forms the skeleton of the neck and are fourteen in number. They have long rod like bodies. The cranial extremity of each vertebra is concave from side to side and convex from above downward, while the caudal extremity presents the reverse condition—i.e., convex from side to side and concave from above downward—which allows a great degree of movement in this region. The transverse processes arise from below the cranial articular process and are prolonged behind by spicules of bone—the styloid processes. These represent cervical ribs projecting in a caudal direction. The foramen transversarium is present at the root of each transverse process. The spinous process are present , except towards the end of the series, poorly developed spinous processes. The ventral process is of considerable size in the last four or five bones (Fig. 4.7).

The first cervical vertebra or atlas, differs from all the rest in its small size and its narrow ring-like form. The ventral arch possesses cranially a deep concave articular area for articulation with the single occipital condyle of the skull. Caudally, there are three facets for articulation with the axis – one median facet on the ventral arch articulates with odontoid process and two facets on the caudo-lateral aspects of the dorsal arch to articulates with the cranial articular process of the axis.

The second vertebra or epistropheus or axis is relatively short, and distinguished by the presence of a small process at the cranial end of the body, the dens, which, passes through the ring of atlas and reaches upto the occipital condyle and articulates with it. The cranial articular processes are small and they articulate with the facets on the caudal aspects of the dorsal arch of the atlas in fowl. The dorsal and ventral spines are present and transverse processes are absent.

Thoracic Vertebrae

There are seven thoracic vertebrae of the fowl which carry ribs. The first and sixth are free, while the second to fifth are fused into one piece to form Notarium. The spinous and ventral processes of second to the fifth being connected into prominent plate-like ridges. Thin bony plate fills the gaps between their transverse processes. The seventh thoracic get fused to the lumbo-sacral segment. Each vertebra has a strong, but relatively short body, which, possesses a well-developed ventral process except in the last thoracic. The transverse processes are prominent and have smooth tubercular facets at their extremities for articulation with ribs (Fig. 4.7).

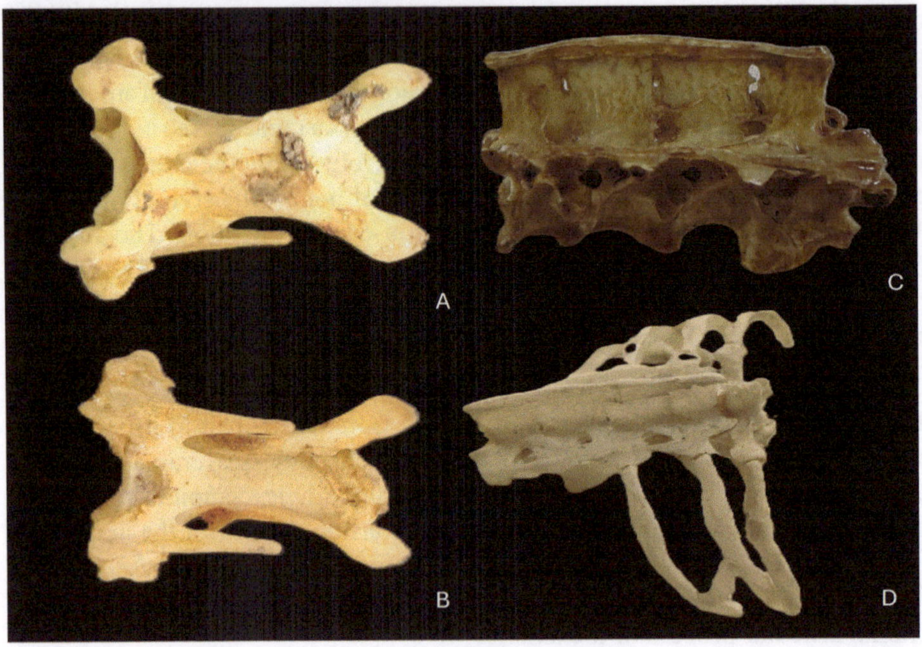

Fig. 47: Photograph showing the dorsal aspect (A) and ventral aspect (B) of cervical vertebra and thoracic vertebrae (C), along with ribs (D) of fowl

Lumbo Sacral Mass

The lumbar and sacrum are fused and called lumbosacral mass in fowl. They are fourteen in number. This lumbosacral mass fused with the seventh thoracic and the first coccygeal (altogether 16) and form one mass called the synsacrum. This forms a rhomboid mass which is present between the two pelvic bones. The spines form a crest in the cranial third but are absent caudally. The transverse ridges on the ventral face indicate the positions of the transverse processes. The extremities of all the transverse processes fuse with the medial border of the ilium.

Coccygeal Vertebra

They are seven in number in fowl. The first coccygeal vertebra is fused with the lumbo sacrals. the last coccygeal is the largest and pyramid in shape. It is formed by the union of several vertebrae, is known as the pygostyl. It forms a foundation for the feathers of the tail. The intermediate ones are typical vertebrae. The transverse processes of these are well developed, and the spines are bifurcated.

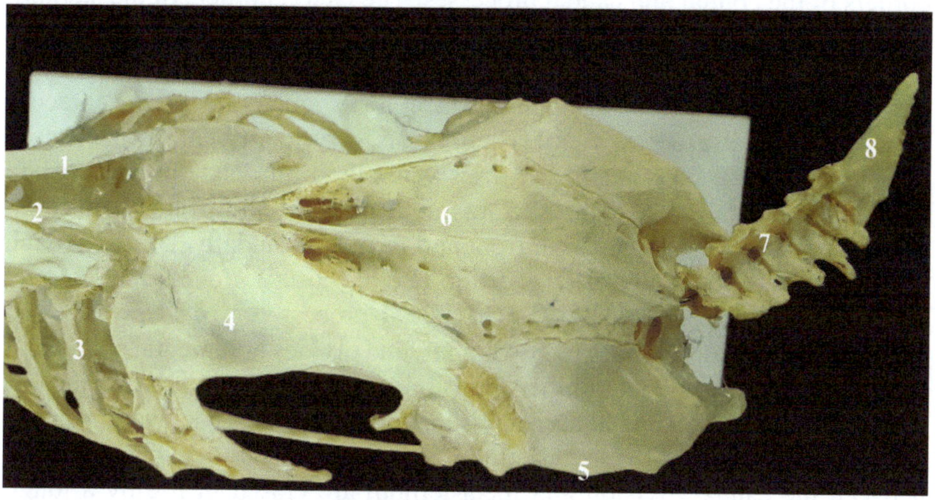

Fig. 4.8: Photograph Showing the Synsacrum of Fowl

1. Scapula
2. Notarium
3. Ribs
4. Ilium
5. Ischium
6. Synsacrum
7. Coccygeal
8. Pygostyle

Ribs

There are 7 pairs of ribs in fowl. Each rib is made up of two segments, an upper vertebral rib and a lower sternal rib. The first, second, and sometimes the seventh are floating ribs and do not have sternal part. The proximal end of each rib consists of a head, neck, and a tubercle. The head of each rib articulates with the capitular cavity of a single thoracic vertebra and the tubercle with the transverse process of a vertebrae. Except for the first and seventh ribs, the posterior border of each vertebral rib presents an uncinate process directed backward to overlap the succeeding rib and connected therewith by a ligament. The distal end of the sternal segment has an articular eminence which fits into a depression on the edge of the sternum (Fig. 4.2).

Sternum

The sternum is a quadrilateral, curved plate with processes projecting from each angle and from the middle of the cranial and caudal borders. It is formed by the fusion of five pieces, which have separate centers of ossification in the chick. The body of the bone is the central wide part. The dorsal surface of the body is concave and pierced by foramina, which permit air to enter the interior from the adjacent air sacs. The ventral surface is encroached upon by the lower borders of the cranial and caudal processes. The lateral borders of the sternum are pitted by four or five depressions into which the sternal segments of the ribs are received. The cranial process, or rostrum or manubrium, is short, and on either side of its root is an elongated facet for articulation with the coracoid bone. The caudal process, or metasternum, is very long, being in fact more extensive than the body itself. It carries ventrally a thin plate of bone called the keel or sternal crest, which affords attachment to the pectoral muscles. The sternal crest/ keel is broad in front and fades out caudally. United to each lateral border of the body are the cranio-lateral (costal) process in front and the caudo-lateral process behind. The costal process projects upwards and forwards. The caudo-lateral process is short and divides into medial and lateral xiphoid processes. The lateral division is broad, plate-like, and covers the last two or three ribs. The medial division is longer and projects backward, parallel to the lateral border of the metasternum. The deep notches between them and between the caudal division and metasternum are closed in life by a tough membrane. The plate-like process of bone (the caudal lateral process) which projects from the caudal angles of the sternum soon divides into two parts. The shorter of these (the oblique process) broadens towards its free end and covers the sternal segments of the last two ribs.

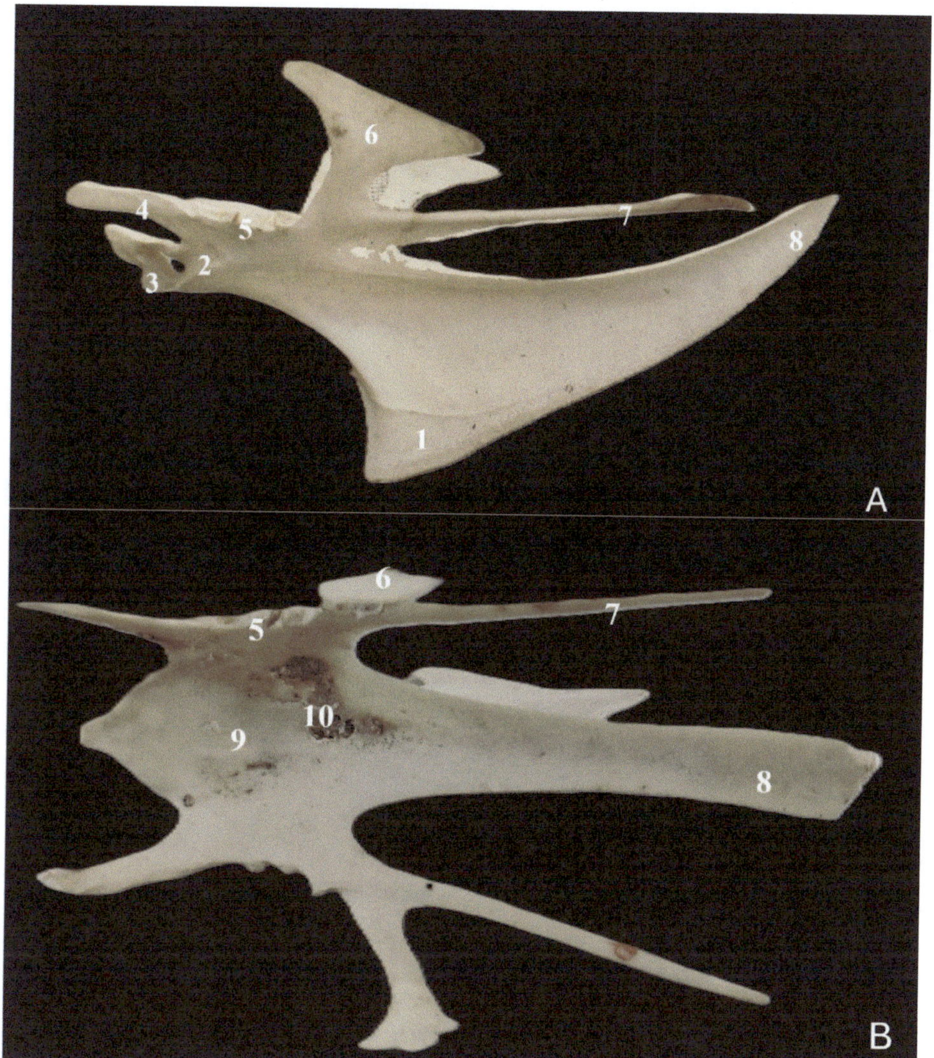

Fig. 4.9: Photograph showingthe Sternum of the Fowl A. lateral view, B. dorsalview: 1. keel, 2. Articular facet for coracoid bone, 3. Rostrum, 4. Cranio- lateral process, 5. Articular facet for ribs, 6. thoracic process, 7. Caudo lateral process, 8. Metasternum, 9. Body, 10. Pneumatic foramen

Index

V

W

X